Yuri Dubrova

Status of the Dosimetry for the Radiation Effects Research Foundation (DS86)

Committee on Dosimetry for the Radiation Effects Research Foundation

Board on Radiation Effects Research

Division on Earth and Life Studies

National Research Council

NATIONAL ACADEMY PRESS
Washington, D.C.

NATIONAL ACADEMY PRESS • 2101 Constitution Avenue, NW • Washington, DC 20418

NOTICE: The project that is the subject of this report was approved by the Governing Board of the National Research Council, whose members are drawn from the councils of the National Academy of Sciences, the National Academy of Engineering, and the Institute of Medicine. The members of the committee responsible for the report were chosen for their special competences and with regard for appropriate balance.

This report was prepared under Department of Energy Cooperative Agreement DE-FC03-97SF21318 between the National Academy of Sciences and the Department of Energy.

International Standard Book Number: 0-309-07559-9
Library of Congress Control Number: 2001012345

Additional copies are available from:
National Academy Press
2101 Constitution Ave., NW
Box 285
Washington, DC 20055
800-624-6242
202-334-3313 (in the Washington Metropolitan area)
http://www.nop.edu

Printed in the United States of America

THE NATIONAL ACADEMIES

Advisers to the Nation on Science, Engineering, and Medicine

National Academy of Sciences
National Academy of Engineering
Institute of Medicine
National Research Council

The **National Academy of Sciences** is a private, nonprofit, self-perpetuating society of distinguished scholars engaged in scientific and engineering research, dedicated to the furtherance of science and technology and to their use for the general welfare. Upon the authority of the charter granted to it by the Congress in 1863, the Academy has a mandate that requires it to advise the federal government on scientific and technical matters. Dr. Bruce M. Alberts is president of the National Academy of Sciences.

The **National Academy of Engineering** was established in 1964, under the charter of the National Academy of Sciences, as a parallel organization of outstanding engineers. It is autonomous in its administration and in the selection of its members, sharing with the National Academy of Sciences the responsibility for advising the federal government. The National Academy of Engineering also sponsors engineering programs aimed at meeting national needs, encourages education and research, and recognizes the superior achievements of engineers. Dr. William A. Wulf is president of the National Academy of Engineering.

The **Institute of Medicine** was established in 1970 by the National Academy of Sciences to secure the services of eminent members of appropriate professions in the examination of policy matters pertaining to the health of the public. The Institute acts under the responsibility given to the National Academy of Sciences by its congressional charter to be an adviser to the federal government and, upon its own initiative, to identify issues of medical care, research, and education. Dr. Kenneth I. Shine is president of the Institute of Medicine.

The **National Research Council** was organized by the National Academy of Sciences in 1916 to associate the broad community of science and technology with the Academy's purposes of furthering knowledge and advising the federal government. Functioning in accordance with general policies determined by the Academy, the Council has become the principal operating agency of both the National Academy of Sciences and the National Academy of Engineering in providing services to the government, the public, and the scientific and engineering communities. The Council is administered jointly by both the Academies and the Institute of Medicine. Dr. Bruce M. Alberts and Dr. William A. Wulf are chairman and vice chairman, respectively, of the National Research Council.

Preface

The dosimetry of the atomic-bomb survivors of Hiroshima and Nagasaki has been a subject of great importance for several decades to the individual survivors, whose estimated dose depends on it, and second to *all* the peoples of the world, because estimates of risk of possible late effects (especially cancer) of an exposure to ionizing radiation are based mainly on the studies of the survivors. The continuing studies of the A-bomb survivors are the most complete and sophisticated evaluations of health effects in an irradiated population that are available to us. Knowledge of the dosimetry has increased because new techniques of calculation of radiation transport and new techniques of measurement continue to be developed and applied. These can improve our knowledge of the dosimetric circumstances of exposure to the A-bombs.

In 1986, the senior committees of the United States (F. Seitz, Chair) and Japan (E. Tajima, Chair) approved the adoption of a new dosimetry system, DS86, for use by the Radiation Effects Research Foundation (RERF) in reconstructing the doses to the survivors. At the time, DS86 was believed to be the best available; it replaced the previous tentative system (T65D) completely, and it has shown itself to be superior to any previous system. It is also the first system to have direct experimental confirmation of the main component of the dose to the organs of exposed people—gamma rays. Nevertheless, questions have arisen about some features of DS86, in particular, whether it adequately describes the neutrons released by the Hiroshima bomb. Measurements of thermal-neutron activation in some materials have suggested more fast neutrons at greater distances from the hypocenter than calculated in DS86. This unresolved discrepancy has tended to cast suspicion on the validity of DS86 dose estimates for dose specifications for the survivors and as a basis of estimation of risk of expected effects in other exposed persons.

For these reasons, the Committee on Dosimetry for RERF, which was set up by the National Research Council (NRC) more than a decade ago at the request of the US Department of Energy, has written this report to describe the present status of DS86 and to recommend studies needed for a possible further-improved dosimetry system. After studying the problems over a period of years, the committee convened a public forum to discuss extant difficulties with DS86 in Irvine, CA in 1996, at which a number of Japanese colleagues were present. During the meeting, a number of actions that could improve DS86 were discussed and were used as the basis of the committee's subsequent letter report, which included a number of recommendations. Some of the recommendations already have been partially implemented; others are being pursued vigorously by the US and Japanese scientists who have continued to study the problem in their laboratories on the advice of this committee and the Japanese senior dosimetry committee. Formal scientific working groups have been set up recently in the United States and in Japan to undertake and complete these studies and two members of the Committee on Dosimetry for the RERF, R. Christy and R. Young, have agreed to step down from the committee to serve on the new working groups and to assist in the implementation of this report's recommendations.

Concerns within the radiation protection and risk assessment communities, the findings of the NRC committee, the commitment of the US Department of Energy and the Japanese Ministry of Health, Labour and Welfare to complete medical follow-up studies of the A-bomb survivors, and recent technical developments have all come together to set the stage for the current reassessment effort. It is now the hope of both Japanese and US scientists that an improvement to DS86 can soon be described in the form of a revised dosimetry system and approved for adoption. It is not expected that a revised system will differ greatly from DS86.

In the meantime, the current report, which has been prepared by the committee members, will describe the present status of DS86, some of the apparent discrepancies that are now being investigated, some of the approaches recommended to solve the problems, and some results already achieved.

WARREN K. SINCLAIR, *Chairman*
Committee on Dosimetry for the RERF

Acknowledgments

The committee members and the National Research Council are especially appreciative for the valuable input provided by numerous members of the Japanese and US Dosimetry Working Groups. Many of these scientists attended meetings in Japan and the United States and presented their data for consideration by the members of this committee. Their work is also represented in the references discussed and listed in this report. In addition, several scientists from Japan and the US hosted Wayne Lowder, Takashi Maruyama, and Harry Cullings, who visited their laboratories to review data related to the committee's work. We wish to thank those scientists who were willing to share their data with the committee and who provided data for the committee's analysis. A special thanks is extended to scientists Harry Cullings and Takashi Maruyama, whose dedicated survey and analysis of the measurement data was critical to the work of this committee.

The committee members are especially appreciative of the administrative assistance, logistical arrangements, and assistance with the manuscript preparation provided by Doris E. Taylor, Cathie Berkley, Benjamin Hamlin and Eric Truett.

This report has been reviewed in draft form by persons chosen for their diverse perspectives and technical expertise, in accordance with procedures approved by the National Research Council's Report Review Committee. The purpose of this independent review is to provide candid and critical comments that will assist the institution in making the published report as sound as possible and to ensure that the report meets institutional standards for objectivity, evidence, and responsiveness to the study charge. The review comments and draft manuscript remain confidential to protect the integrity of the deliberative process. We wish to thank the following persons for their participation in the review of this report:

David Auton, Defense Nuclear Agency (retired)
Randall S. Caswell, National Institute of Science and Technology (retired)
Edward Epp, Harvard University (retired)

Daniel O. Stram, University of Southern California, Los Angeles
Paul Goldhagen, Department of Energy, Environmental Measurements
    Laboratory, New York
George Kerr, Oak Ridge National Laboratory (retired)
Atsuyuki Suzuki, Quantum School of Engineering, University of Tokyo,
    Tokyo, Japan
Francis X. Masse, Massachusetts Institute of Technology, Cambridge

Although the reviewers listed above have provided many constructive comments and suggestions, they were not asked to endorse the conclusions or recommendations nor did they see the final draft of the report before its release. The review of this report was overseen by David G. Hoel, Medical University of South Carolina and John Dowling, Harvard University. Appointed by the National Research Council, they were responsible for making certain that an independent examination of this report was carried out in accordance with institutional procedures and that all review comments were carefully considered. Responsibility for the final content of this report rests entirely with the authoring committee and the institution.

Contents

Executive Summary

The system now used for estimating radiation doses to individual survivors of the atomic bombs is called DS86 (for Dosimetry System 1986). It was introduced in 1986 as the first comprehensive system to provide dose estimates for the nearly 100,000 survivors being studied by the Radiation Effects Research Foundation (RERF). It replaced an earlier system T65D (Tentative 1965 Dosimetry). The main component of the radiation dose to organs of exposed people (over 98% of the absorbed dose) is gamma radiation, and DS86 gamma-ray calculations have been verified by direct experimental measurements using thermoluminescence. DS86 estimates the dose to each of several organs of a given survivor, allowing for his or her shielding by house, terrain, and his or her own tissues. Major components of the radiation field include prompt and delayed neutrons as well as early and late gamma rays (Roesch 1987).

Uncertainty in DS86 estimates of organ dose was provisionally discussed by Roesch (1987) and more comprehensively by Kaul and Egbert in 1989, but a complete evaluation of uncertainty in all aspects of DS86 is still needed. One important uncertainty in DS86 concerned the neutron component. Measurements of cobalt-60 (^{60}Co) and europium-152 (^{152}Eu) activated by thermal neutrons suggested that there were more fast neutrons at great distances (>1500 m) than DS86 had indicated. The discrepancy became greater in Hiroshima when chlorine-36 (^{36}Cl) activation measurements were added and apparently became smaller or nonexistent in Nagasaki when more detailed calculations were made. The DS86-calculated neutron component of the dose in Hiroshima is only 1–2% of the total organ absorbed dose, and that in Nagasaki only one-third of this, so the impact of a discrepancy in the neutron component is not necessarily large, depending on the choice of relative biological effectiveness (RBE) for the neutrons.

1

The National Research Council's Committee on Dosimetry for the Radiation Effects Research Foundation was formed at the request of the Department of Energy (DOE) soon after DS86 was introduced. The committee was charged with monitoring the status of DS86 and assessing its relevance in light of any new evidence. It has continued to be both the repository and the forum for discussion of revisions of DS86. In a 1996 letter report to DOE, the committee recommended a program to help solve the neutron-discrepancy problem through direct measurement of activation of nickel-63 (^{63}Ni) due to release of fast neutrons from both bombs. That program and a program to evaluate thermal-neutron measurements (also stimulated by the committee) are still in progress.

Cooperation between dosimetry working groups from Japan and the United States and committees from both countries has been fostered by joint meetings like those in 1996 in Irvine and in March 2000 in Hiroshima, the latter supported largely by Japan's Ministry of Health and Welfare. Both countries, on the advice of these committees, have set up formal working groups to pursue issues related to DS86 and to deliver an updated dosimetry for RERF.

This report describes the status of DS86—specifically, the discrepancies that should be investigated, some of the approaches recommended to remedy them, and some preliminary results.

GAMMA-RAY MEASUREMENTS

When DS86 was adopted, the main component of the dose, gamma rays, had been directly measured by thermoluminescence in quartz that was inside bricks and tiles in structures up to about 2 km from the hypocenter in both cities. These and other measurements since 1986 are described in Chapter 2, which discusses and considers the magnitude of various sources of uncertainty in the measurements, including fading, calibration, energy response, and background. The measurements are slightly higher than calculations in some regions and lower in others. The uncertainty in the gamma-ray fluences measured and calculated is around ±20% (well within the range of uncertainty to be expected in thermoluminescence measurements), but arbitrarily reducing the measurements by 20% does not improve agreement with DS86 calculations.

NEUTRON MEASUREMENTS

As noted above, discrepancies have been reported in fast-neutron calculations of fluence and dose compared with measured thermal-neutron activation of ^{60}Co, ^{152}Eu, and ^{36}Cl; the ratio of measured to calculated fluence can reach 10:1 at 1500 m in Hiroshima and higher at greater distances. Discrepancies in Nagasaki seem small or nonexistent. The latter finding suggests a problem with the Hiroshima source term. Attempts to model a new source that fits the measurement data have not resulted in a plausible source. Thermal neutrons result from fast neutrons but do not contribute significantly to the dose as fast neutrons do. Consequently, direct measurements of fast neutrons at various distances from the epicenter would constitute a superior test of the nature and magnitude of the problem. A method involving fast-neutron activation of copper to nickel—$^{63}Cu(n,p)^{63}Ni$—has been proposed and is now in use. The ^{63}Ni is being measured in Japan on the basis of its radioactivity and in the United States and Germany with accelerator mass spectrometry of copper samples from known locations in Hiroshima. Preliminary results are available but have not yet been fully evaluated. Samples from Nagasaki will, hopefully, also be available for measurement.

On the recommendation of this committee in its 1996 letter report, a joint US-Japan team of experienced measurement personnel (T. Maruyama, W. Lowder, and, later, H. Cullings and H. Beck) was set up to reexamine all aspects of all gamma and neutron measurements with regard to how uncertainties are represented and what is included in them. Particularly important are how background is measured and subtracted and how data are selected at low activity levels. In some cases, background is greater than the sample signal.

REEVALUATION OF DS86

Since 1986, many revisions in the parameters of DS86 have been proposed (but not incorporated) that would improve the calculations. These have included changes in transport cross sections and transport codes and refinement of the calculations (increases in numbers of gamma-ray and neutron energy groups utilized). The changes, when tested, greatly improved the agreement with thermal-neutron measurements in Nagasaki, essentially removing the discrepancies, but the major discrepancy in Hiroshima beyond 1200 m remained.

Many scenarios for the Hiroshima bomb were explored to try to understand the problem. It was found that no addition of fast (more penetrating) neutrons, either from leakage through a cracked bomb casing, or from an alternative plausible source term, could account for the increase in thermal-neutron activation and still agree with the well-known fast-neutron activation measurements of sulfur-32 (^{32}S) made in situ soon after the bomb explosion.

BIOLOGICAL DOSIMETRY

Biological dosimetry is not generally expected to achieve the detailed precision of good physical measurements, but biological assays can provide a sound perspective on whether the physical dosimetry has led to reasonable results. They can be used to test the dosimetry system and provide evidence to confirm DS86 or to suggest the presence of a problem.

In the RERF research program in Hiroshima and Nagasaki, biological methods of evaluation of doses have been especially important. For example, cancer-risk estimates based on epidemiological data have been about twice as great for the

Hiroshima survivors as for the Nagasaki survivors. Many uncertainties are associated with these estimates for both cities, and there are many possible reasons for the differences between them, including the much more complex terrain-shielding problems in Nagasaki; but one possible reason for the differences is the dosimetry itself. The differences might be attributable, at least in part, to the greater number of neutrons in Hiroshima than in Nagasaki. DS86 predicts that the neutron organ doses are 3 times greater in Hiroshima than in Nagasaki. As noted previously, DS86 might substantially underestimate the neutron fluence in Hiroshima, to judge from preliminary measurements.

Two techniques have been used for direct biological measurement. Measurements of stable chromosomal aberrations can be compared with DS86 dose calculations but these aberration measurements show considerable scatter, indicating only a rough correspondence with DS86 estimates. Measuring electron-spin resonance in tooth samples has been possible in about 60 survivors and yields estimates in good agreement with estimates based on chromosomal aberrations for the same individuals. Within broad uncertainty limits, however, both methods yield results consistent with DS86 estimates for the same people, except for the Nagasaki factory workers.

UNCERTAINTY

In estimating doses, it is essential to consider sources of and contributions to uncertainty. DS86 included a preliminary assessment of these uncertainties. The assessment was based on fractional standard deviations estimated for various key parameters and on nonparametric methods derived mainly from the judgment of the DS86 authors, with little analytical support. The National Research Council's initial review of DS86 recommended a rigorous uncertainty analysis that would use improved uncertainty-input values for each aspect of the dosimetry system (NRC 1987). A report produced in response to that recommendation, indicated fractional standard deviations of around 25–40% (Kaul and Egbert 1989). The present report indicates (in Chapters 2 and 3) that uncertainties might be larger when all relevant factors are taken into account, item by item, for both gamma rays and neutrons. Uncertainty analysis has become more feasible because of the availability of new information on possible sources of uncertainty and the availability of faster computers, which permit benchmark and sensitivity studies.

IMPLICATIONS FOR RISK ASSESSMENT

One potentially important aspect of the possibly greater neutron fluence in Hiroshima is whether such neutrons have an important effect on estimates of risk posed by gamma rays, which are the main component of the dose and therefore the primary cause of late effects, such as cancer. It is particularly important because DS86 considers neutrons to be a small component that might be ignored with regard to their tumorigenic effect. But neutrons might still be significant in DS86, in view of their potentially large relative biological effectiveness (RBE). In the illustration given in Chapter 7, the DS86-estimated neutron fluence with reasonable RBEs of 20–50 at low doses might cause 23–43% of the effect in Hiroshima, leaving 57–77% for gamma rays. In Nagasaki, the corresponding proportion of neutron effect is 9–20%. The effects would therefore be expected to be 1.2–1.4 times greater in Hiroshima because of the extra neutrons in Hiroshima alone, and indeed that is roughly the case. However, so many other complex factors are involved in differences between the two cities, including terrain and shielding, that it is highly questionable whether that difference can be attributed solely to the neutrons.

If the neutron component is actually larger in Hiroshima than in DS86—for example, 3 times larger at 1500 m—it would imply an even greater proportion of the effect for neutrons and less for gamma rays. In any case, the gamma-ray risk might be lower than previously estimated because of the contribution of the neutrons at their present level in DS86 in Hiroshima (see illustration in Chapter 7). The gamma-ray risk (i.e. lifetime attributable cancer mortality risk) might drop from the present 5–6% Sv^{-1} to 4–5% Sv^{-1}—still well within the current range of uncertainty of these derived values. If more neutrons were present at Hiroshima, the effect could be to reduce the gamma-ray risk estimates somewhat further.

CONCLUSIONS AND RECOMMENDATIONS

Although DS86 is a good system for specifying dose to the survivors and for assessing risk, it needs to be updated and revised. Uncertainties have not been fully evaluated and might amount to more than the 25–40% in fractional standard deviations of parameters (Kaul and Egbert 1989). While the calculated gamma-ray fluences agree well with values measured with thermoluminescence and constitute the main component of the dose to the survivors, more work needs to be done to establish the magnitude of the neutron component and to assess the extent to which the neutron component (small in DS86) affects (lowers) the estimates of gamma-ray risk.

The committee offers the following recommendations regarding the revision of DS86 that is clearly needed and that hopefully will be completed in 2002:

- The present program of [63]Ni measurements should be pursued to completion.
- All thermal-neutron activation measurements, particularly those with [36]Cl and [152]Eu, should be reevaluated with regard to uncertainties and systematic errors, especially background (see Chapter 3).
- Critical efforts to understand the full releases from the Hiroshima bomb by Monte Carlo methods should be continued.
- Adjoint methods of calculation (i.e., going back from the field situation to the source term) should be pursued to see whether they help solve the neutron problem.

- Local shielding and local-terrain problems should be resolved.
- The various parameters of the Hiroshima explosion available for adjustment including the height of burst and yield, should be reconsidered in the light of all current evidence in order to make the revised system as complete as possible.
- A complete evaluation of uncertainty in all stages of the revised dosimetry system should be undertaken and become an integral part of the new system.
- The impact of the neutron contribution on gamma-ray risk estimates in the new system should be determined.

1

Introduction

In March 1986, at a meeting in Hiroshima, the senior dosimetry committees of the United States (Chair F. Seitz) and Japan (Chair E. Tajima) approved the adoption of a new dosimetry system, DS86 (Dosimetry System 1986), for use in estimating the doses received by people exposed to the Hiroshima and Nagasaki atomic bombs in August 1945. The dosimetry committees and those who worked with them (in the U.S. and Japanese working groups, chaired by R. Christy and E. Tajima, respectively) were at the time well aware of the critical importance of the dosimetry system, not only for the reconstruction of the doses to the survivors, but also for estimating the risk of late cancer effects in the survivors and possibly, as noted more recently (Shimizu and others 1999), the risk of some noncancer health effects. Those estimates depend critically on the dosimetry. The estimates of risk are the main basis of our knowledge of the late health effects to be expected in any future population exposed to substantial doses of ionizing radiation. They are therefore of great importance to the people of the entire world. The risk estimates also are the main basis of recommended radiation-protection dose limits for radiation workers and for the public after inadvertent exposure to ionizing radiation (ICRP 1991; NCRP 1993).

DS86

DS86 was the first fully comprehensive computerized dosimetry system to be recommended for use with the atomic-bomb survivors. It replaced the former T65D system with state-of-the-art knowledge of all known measures related to the Hiroshima and Nagasaki explosions. DS86 incorporates computations and models that describe the yield and radiation output of the bombs, the free-field radiation enviror ment, the shielding circumstances of the survivors, and the body shielding of the various organs. It is a modular system with separate databases for each of the free-field radiation components, for each of several distinct shielding environments, and for each of many different organs.

The free-field components include prompt neutrons, early gamma rays (prompt-fission gamma rays and gamma rays from inelastic scattering and capture of prompt neutrons), late gamma rays (from fission products and from delayed neutrons), and delayed neutrons. A new or revised treatment of any of those components can readily be introduced by appropriately substituting a new database for an existing one. The shielding databases include models for all survivors with nine-parameter shielding and all survivors with globe-data shielding descriptions (Roesch 1987).

Uncertainties were estimated for the shielding and organ environments by calculating fractional standard deviations among the many shielding and phantom environments that had been computed. They were combined with uncertainties in the free-field radiation fluences to provide a preliminary estimate of uncertainty in the computed doses. Uncertainty evaluation was incomplete at the time of the adoption of DS86 (Roesch 1987). Later, Kaul and Egbert (1989) presented the US dosimetry committee with a draft of a preliminary uncertainty analysis; this analysis was revised in 1992 but is still regarded as preliminary.

GAMMA RAYS

In the DS86 calculations of kerma and dose to exposed people, the codes and data used were superior to any used previously. Especially for doses to organs deep in the body, gamma rays dominate, amounting in Hiroshima to 98–99% of the total absorbed dose. In Nagasaki, where the neutron fluence at a specified total dose is only about one-third that in Hiroshima, the percentage contribution of neutrons is even lower.

Not only were the calculations of gamma rays believed to be improved in DS86, but a most important consideration was the experimental confirmation of gamma-ray doses by measurements (with thermoluminescent dosimetry or TLD) of the gamma-ray signal in quartz, brick, and tile samples in both cities. Agreement between measurement and calculation is quite good over a wide range of distances from the hypocenter in both cities. That bears repeating for emphasis: the most important component of the dose, gamma rays, is experimentally well verified (see Chapter 2).

NEUTRON MEASUREMENTS AND DISCREPANCY

The contribution of fast neutrons to the dose in organs deep in the body is estimated in DS86 at around 1–2% of the total absorbed dose in Hiroshima in the dose range of about 0.5–2 Gy. Nevertheless, especially because of the potentially high relative biological effectiveness (RBE) of neutrons compared with gamma rays, the neutron component, although small, might be a contributor to the effects of ionizing radiation from the atomic bombs.

It was known by the dosimetry committees and the working groups (Roesch 1987) at the time (in 1986) that DS86 contained possible flaws. In particular, questions had arisen because some induced radionuclides, such as ^{60}Co, activated by thermal neutrons in steel structures (Hashizume and others 1967; Loewe 1984; Roesch 1987), presented problems that at that time were unresolved. Notably, they seemed to indicate the possibility that at increasing distances from the hypocenter of both explosions the numbers of thermal neutrons exceeded those predicted by the DS86 calculations. Eventually because these measurements could not be easily or quickly verified or explained, it was decided to proceed with the application of DS86 to the survivors, since DS86 appeared superior to predecessor systems, and the corrections involved in the new system needed to be implemented without delay. The impact of the changes resulting from the application of DS86 on the risk estimates for cancer (NRC 1990; UNSCEAR 1988) and on the recommendations for radiation protection (ICRP 1991; NCRP 1993; NRPB 1993) is now well known. With other factors that were included in the reassessment, the changes contributed to the increase (by about a factor of 3) in the risk estimates for occupational exposures and exposures of the public.

The apparent discrepancies between calculation and measurement of thermal-neutron activation first noted with ^{60}Co were emphasized by measurements of [152]Eu in Japan (Nakanishi and others 1983; Okajima and Miyajima 1983) and are discussed in the DS86 report (Roesch 1987). More recent measurements of [36]Cl in concrete in Hiroshima (Straume and others 1992) appear to be consistent with those results; these [36]Cl measurements culminated in a review of all the activation measurements known at the time (Straume and others 1992). The review further emphasized the higher measured values at large distances in Hiroshima (a factor of about 10 between measured and calculated values at a 1500 m slant range and still higher at longer distances). In Nagasaki, after revised calculations of neutron fluences and additional [36]Cl measurements, the discrepancy appeared smaller and possibly nonexistent (Straume and others 1994). If the measurements were indeed correct for both cities, the problem might well rest in the physical characteristics of the source term of the Hiroshima bomb rather than in transport or other possible problems between source and detector. Possible new source terms for the Hiroshima bomb have been vigorously pursued and are still being considered, but no new and plausible source term that fits all the data has yet been proposed.

Even though the fast-neutron component of the absorbed dose is small, with a high RBE (20–50 at the low neutron doses in question), the neutron equivalent dose in Hiroshima might not be negligible. Any increase in fast-neutron fluence implied by the thermal-activation measurements could therefore have an important effect on the total equivalent dose in Hiroshima and correspondingly on the risk estimates for the Hiroshima portion (two-thirds) of the survivor population (see Chapter 7). Such uncertainties in the neutron contribution to the absorbed dose have also cast broad suspicion on DS86 for assessing the doses and the risks to the survivors and for deriving the gamma-ray risk estimates from the epidemiological studies at the Radiation Effects Research Foundation (RERF). With regard to the survivors themselves the concerns are largely unfounded because the effects of the actual radiation in Hiroshima and Nagasaki, including its neutron component, have been directly observed and are known without need to separate the effects of the gamma rays and the neutrons. Uncertainties in the neutron dose can however be important to the derivation of the risk due to the gamma rays. It has, however, been pointed out (NCRP 1997; Pierce and others 1996; Preston and others 1993) that the impact on risk estimates probably could not exceed 20%, because those estimates are derived mainly from comparatively high doses (such as 1 Gy and higher, which correspond to ranges at most 1200 m from the hypocenter) and because the discrepancies occur only in Hiroshima, apparently not in Nagasaki. Other workers (Kellerer and Nekolla 1997; Rossi and Zaider 1990; Rossi and Zaider 1996; Rossi and Zaider 1997) point out, however, that at low doses, the health effects, if any (depending on the dose-response relationship in that region), might be due to a significant extent to the neutrons in Hiroshima and not to gamma rays, for which the risk estimates are derived. Again, Nagasaki seems not to be subject to this uncertainty, but risk estimates for Nagasaki alone are less certain, because the sample is only about half the size of the Hiroshima sample and because of other factors, especially related to shielding and terrain, than risk estimates for the combined Hiroshima-Nagasaki sample. In 1996 (Pierce and others 1996), estimates of risk based on Nagasaki alone were substantially lower than those based on the combined sample. In any event, the idea has persisted in some quarters that DS86 is in question and needs to be reexamined and amended or replaced if necessary. At the very least, the "neutron problem" has contributed to the greater uncertainties now recognized in the risk estimates derived from this important source (NCRP 1997).

A fuller understanding of the features of the Hiroshima explosion—which, unlike the Nagasaki explosion, has no counterpart for comparison among other (test) weapons—would obviously be highly desirable but might not in fact be possible. In the meantime, it would clearly be important to be able to measure fast neutrons at distances from the hypocenter in Hiroshima directly and to establish better the magnitude, if not the explanation, of the thermal-neutron discrepancy. It would also be highly desirable to measure fast neutrons at Nagasaki to confirm that a similar discrepancy does not exist there.

During the time since DS86 was introduced, it became evident (Ruehm and others 2000b; Shibata and others, 1994) that a method to assay fast neutrons from the bombs directly by using ^{63}Ni from an (n,p) reaction in ^{63}Cu was feasible either with radioactivity measurement (T $\frac{1}{2}$ = 100y) or with mass spectrometry (Ruehm and others 2000b; Straume and others, 1996).

THE NATIONAL RESEARCH COUNCIL
COMMITTEE ON DOSIMETRY FOR THE RERF

The National Academies' National Research Council had set up the US Senior Dosimetry Committee with Frederick Seitz as chair to oversee the work of the Department of Energy (DOE) working group (R. Christy, chair), which, with the Japanese working group, produced DS86. The National Academies essentially disbanded the Senior Dosimetry Committee in 1987 after the production of the report *An Assessment of the New Dosimetry for A-bomb Survivors* (NRC 1987). Later, it was recognized that there were likely to be continuing problems and issues in dosimetry that required a standing committee on dosimetry for RERF to advise the National Academies and others. A new Committee on Dosimetry for the RERF was set up in 1988, with Alvin Weinberg as chair and Warren Sinclair as vice-chair. The committee undertook such duties as approving a relaxation of the stringent requirements on dosimetric factors for DS86 dosimetry assignment, which enabled the survivors with assigned DS86 doses to be increased from about 76,000 to about 86,000 in a total sample of some 92,000 survivors.

In 1992, with the problem of thermal-neutron activation studies coming to the fore on both sides of the Pacific, Sinclair and Weinberg exchanged roles, and Sinclair became chair. Addressing the neutron problem became the top priority of the committee although other features of DS86 had, in the meantime, been addressed by members of the former working group, such as shielding issues, revisions in yield, height of burst, and new measurements of neutron cross sections for which the committee was the focal point for discussion and review. The thermal-neutron activation studies continued to suggest larger numbers of neutrons in Hiroshima than could be accounted for by DS86.

A meeting of the Committee on Dosimetry for the RERF held at Irvine, CA, on May 22–23, 1996, was attended (by invitation) by five members of the Japanese committee who were working on this problem. Later, a formal letter report from the NRC Committee was sent to DOE (Appendix D, NRC 1996). The letter recommended a number of actions to be taken to lead to renewed confidence in DS86 or to its revision. These included a review of all existing thermal-neutron measurements by a joint US-Japan team, initiation of ^{63}Ni measurements of fast neutrons (based on radioactivity in Japan and mass spectrometry in the United States), revisions in other measures related to the Hiroshima bomb, and a thorough examination of uncertainties in all aspects of DS86.

Many groups have come to recognize the importance of the dosimetry in Hiroshima and Nagasaki in the political arena because it underlies current risk estimation for radiation protection and because standards are based on it. That has led to more US government support for investigations to settle issues about DS86.

The Committee on Dosimetry for the RERF was reconstituted in 1998 (Warren Sinclair, chair) with additional expertise and has since continued to concentrate on solving the neutron problem, acting as an advisory group first for the scattered scientists working on the issue and, as of 2000, for the more cohesive US working group of scientists chaired by Robert Young. The efforts of the working group are expected to result in collaboration with a parallel Japan working group in a report detailing a revision or confirmation of DS86 and including a comprehensive uncertainty evaluation. The present NRC report describes in some detail the current situation with DS86 and recent measurements that give some preliminary indications of results for fast neutrons, and it sets the stage for the subsequent work of the joint US and Japan working groups.

UNITED STATES-JAPAN INTERACTIONS

The Committee on Dosimetry for the RERF has interacted closely over the years with corresponding colleagues in Japan, and interactions have been fostered by personnel at RERF itself, including Dale Preston, Shiochiro Fujita, and, recently, Harry Cullings. Those people have provided an invaluable service in obtaining samples for thermal-neutron and fast-neutron measurements in Japan, the United States, and Germany. Japanese colleagues have also attended some US meetings.

As a result of the good offices of Shigenobu Nagataki, chairman of RERF, and assistance from the Ministry of Health and Welfare, Japan, for travel costs, a joint scientific meeting of Japanese and US workers and committees was held in Hiroshima on March 13-14, 2000. The meeting considered many aspects of DS86, including the recent work in Japan, the United States, and Germany on ^{63}Ni assays in copper samples from Hiroshima (Ruehm and others 2000a). The results so far demonstrate the feasibility of applying these methods to measurements at large slant ranges in Hiroshima. Samples from Nagasaki are not yet available but are being eagerly sought, as are additional samples from Hiroshima. It is clear that appropriate background assessment in both fast-neutron and thermal-neutron measurements, with inevitably small signal-to-noise ratios, will be an important feature of the final measurements. Scientists in the United States and Japan had fruitful exchanges, and the meeting resulted in further dedication among the US and Japanese scientists to accelerate and complete investigations related to the dosimetry.

In the 14 years since DS86 was approved, many other physical characteristics related to DS86 have been reassessed, and some improved values have been proposed. These include carrying out calculations with many more neutron and gamma-ray energy groups; reevaluating neutron cross sections in air, nitrogen, and iron over a broad range of energies; possible revisions in maps (Kaul 2000); and reexamining the effects of various characteristics, such as height of burst, yield, shielding, and organ doses. New determinations have not been agreed on. Nevertheless, even if the fast-neutron and thermal-neutron discrepancy is resolved, it will be necessary to consider (soon) whether possible revisions in parameters (see Chapter 4) should be collectively applied in a revised DS86 and adopted for use in specifications of the dose to each survivor. The US working group will address these problems in its report.

The present report reviews the present status of DS86, the gamma-ray dosimetry, and such dosimetry issues as the thermal-neutron discrepancies between measurement and calculation at various distances in Hiroshima and Nagasaki. It recommends approaches and measurements to bring those issues to closure; that is, it sets the stage for the program of the working groups. It also outlines the changes in various physical characteristics relating to DS86 in the last 14 years and encourages the incorporation of the changes into a revised dosimetry system.

In the succeeding chapters, the report reexamines aspects of measurement and calculation for the most important radiation-field components of the dosimetry: gamma-ray measurements, thermal-neutron and fast-neutron measurements, data-quality assessment, improvement in parameters of DS86, some features of neutron transport, biological dosimetry at RERF, uncertainty in DS86, and the implications of the foregoing for risk assessment.

Four appendixes address the dosimetry database at RERF, the uncertainty analysis of neutron-activation measurements in Hiroshima, and the cosmic-ray neutron contribution to sample activation. Appendix D includes the 1996 letter report of this committee. A glossary and a list of references complete the report.

2

Gamma-Ray Measurements

Gamma rays emitted from the Hiroshima and Nagasaki explosions had two major sources. One was the nitrogen-capture gamma rays that arose from the capture of bomb neutrons when slowed in air to thermal energies to emit roughly 5 MeV gamma rays, which are very penetrating in air. These were proportional to the number of neutrons captured in nitrogen. The second source was the gamma rays emitted by the fission products in the fireball (which were proportional to the bomb yield). These gamma rays were emitted mostly between 1 and 10 s after the explosion. A third, much smaller contribution was prompt gamma rays from the device itself.

THERMOLUMINESCENT MEASUREMENTS
IN BRICK AND TILE

A most important feature of DS86 was that thermoluminescent (TL) measurements in brick and tile were used to verify the calculations of free-in-air kerma (FIA) from these gamma rays. The agreement in 1986 between the TL measurements and DS86 calculations was generally good and within the estimates of uncertainty for the DS86 calculations (Roesch 1987). Quartz, like any material used in thermoluminescent dosimetry, contains quantitative radiation-exposure history from the time of its initial annealing at high temperature. The method of retrieving the exposure history in quartz was based on a protocol first developed by Grogler and others (1960) for dating of ancient pottery fragments. The technique was improved by Ichikawa (1965) and Fleming and Thompson (1970), who used the TL method in archeological studies. Higashimura and others (1963) reported that the exposures and doses from gamma rays in Hiroshima and Nagasaki could be measured by this method. A substantial effort was then mounted to obtain suitable material for reconstructing the dose as a function of distance from the bomb hypo-center. Samples of brick and roof tile were selected at homes, universities, temples, a hospital, and so on according to criteria for quantitative exposure measurements. The technique to obtain the TL data on brick and tile exposure was then developed and carried out with extraordinary care.

Three methods of measurement, with variations, have been used. The first was the high-temperature technique, which is a straightforward measurement of the glow curve (luminescence with heating), from a sample of quartz. The sample is separated from the clay-tile matrix and, irradiated at a known exposure, and the glow curve is read. The main trap for electrons displaced by the gamma-ray expo-sure of interest is a relatively high energy trap of about 7 eV and requires near 500°C heating to erase the signal (by annealing) and up to about 300°C for the glow-curve (TL) measurement. The second, or additive, technique attempts to cor-rect for an initial nonlinearity in the dose-response relationship of quartz by irra-diating multiple samples of the extracted quartz at increasing doses, and correct-ing the measurements of the original sample in the nonlinear or supralinear region. The third, predose, technique and its variations use a lower-energy electron trap in quartz near 110°C, which fades rapidly and is not found in normal samples. The stored-dose information in the higher-energy trap can be explored with sufficient heating of the sample to permit transfer from the high-energy to the low-energy trap and then readout of the low-energy information.

All three techniques have been used on specimens of brick and ceramic tiles in Hiroshima and Nagasaki and provide quantitative measurements of the cumulative gamma-ray dose histories of the quartz crystals in the samples. After conversion to estimates of free-in-air tissue kerma at the specimens' locations at the time of the bombing, these measurements can be compared directly with estimates calculated from DS86. The details of the comparison procedure are shown in Figure 2-1.

The gamma-ray kerma at both Hiroshima and Nagasaki was measured directly with the radiation signal stored in brick or tile within the range of the bomb radia-tion. The measurements show that, within the predictable error, the agreement with the DS86 calculation is satisfactory.

However, there are at least three inherent sources of bias in making measure-ments with the TL method and comparing them with DS86 estimates. These are considered here to show the extent of agreement with the DS86 calculations:

• **Fading.** Fading of the radiation signal with time occurs readily in other TL materials, such as LiF, but quartz is considered to have a very stable high-energy trap. Fading of the signal occurs if heating, after exposure, occurs near the trap temperature region, 400–500°C.

• **Backscatter medium.** The response signal in a sample that is part of a large structure, such as a roof or wall with concomitant backscattered radiation from the surrounding material, is larger than the response to a calibration exposure carried out without backing material.

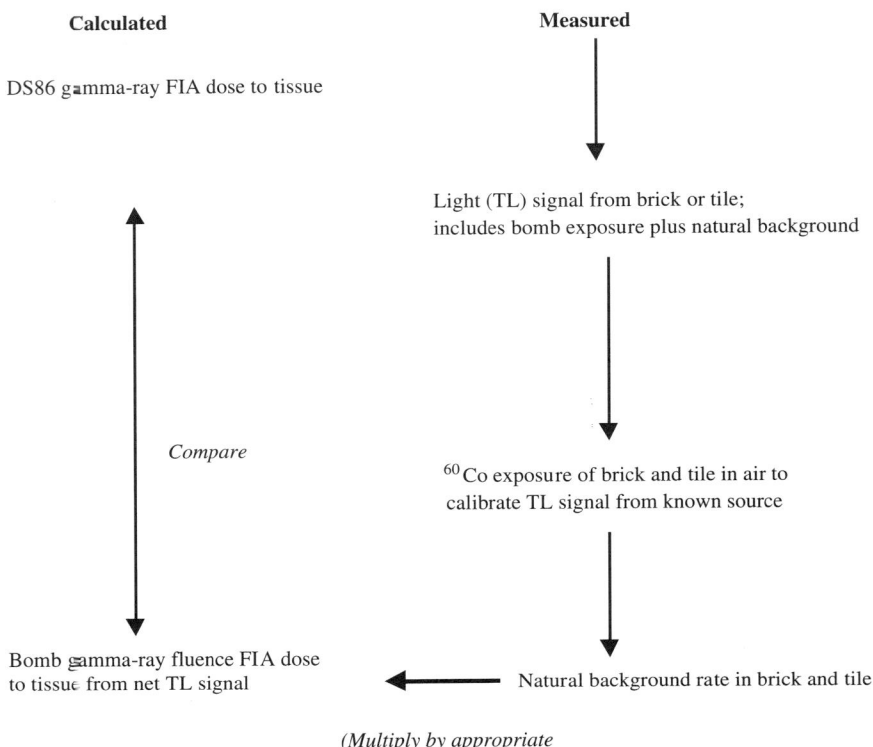

FIGURE 2-1 Measured and calculated quantities and related to gamma dose (free-in-air tissue kerma).

• **Quartz energy response.** At the same air kerma the energy signal in quartz at low photon energies per unit dose (or fluence) is greater than that at higher photon energy.
• **Background issues.** An additional consideration is the uncertainty in the measurement of the natural background radiation signal inherent in any brick or tile, which must be subtracted if bomb radiation is to be measured.

These four factors and their effects on the quality of TL measurements are discussed in more detail in the following sections.

Fading

Figures 2-2 and 2-3, from *The Effects of Nuclear Weapons* (Glasstone and Dolan 1977), show the thermal effects of the Hiroshima bomb. Figure 2-2 shows the bubbling or blistering of common roof tile from the heat of the bomb. The ther-

FIGURE 2-2 Flash marks produced by thermal radiation on asphalt of bridge in Hiroshima. Where the railings served as protection from the radiation, there were no marks; the length and direction of "shadows" indicate the point of the bomb explosion (Glasstone and Dolan 1977).

mal shadow of the railings on the asphalt bridge pavement was used to support the estimates of height of the detonation.

The Effects of Nuclear Weapons provides evidence that blistering of tiles (Figure 2-3) was observed up to 980 m from the hypocenter. The radiant energy at this distance was estimated to be 45 cal/cm^2. Experimental heating of similar tiles showed that an 1800°C impulse for 4 s produced the same blistering, although the heat penetration into the tile was thought to be greater than that in Hiroshima or Nagasaki. No further information on this point is given.

Although temperature of that magnitude would certainly affect the TL signal after exposure, there are no data to confirm fading of a signal when the radiation exposure and the radiant energy occur at nearly the same moment. It is not known

FIGURE 2-3 Blistered surface of roof tile; left portion of the tile was shielded by an overlapping one (0.37 mile from ground zero at Hiroshima) (Glasstone and Dolan 1977).

what depth of penetration the radiant energy would have in a typical brick or tile. If fading due to annealing did occur in samples, the exposures from TL measurements would have been underestimated.

The published studies contain statements suggesting that the tile samples collected had not been substantially heated by the explosion of the bomb or by accidents (Hoshi and others 1987). However, radiant energy as a function of distance from the hypocenter has not been addressed explicitly.

Figure 2-4, also from *The Effects of Nuclear Weapons*, gives estimates of the radian energy at ground distances from a hypothetical hypocenter for different weapon yields for a particular weather condition. Depending on the orientation of the tile with respect to the bomb, there was probably a potential for some annealing of the sample due to high temperatures at least out to 1000 m.

Some of the TL measurements were performed by Edwin Haskell, who indicates that the predose technique used precludes any effect of fading (Haskell 2000). Moreover, if there were fading, the depth-dose curves in tile would not be consistent with the gamma-ray energy attenuation (Haskell 2000).

Overall, the effect of signal fading in the samples measured was considered to be negligible, and there is no reason to disagree with what has been previously concluded.

Backscatter Medium

All calibration exposures were carried out with a ^{60}Co calibration source (E = 1.17, 1.33 MeV) except at the University of Utah, where a ^{137}Cs source

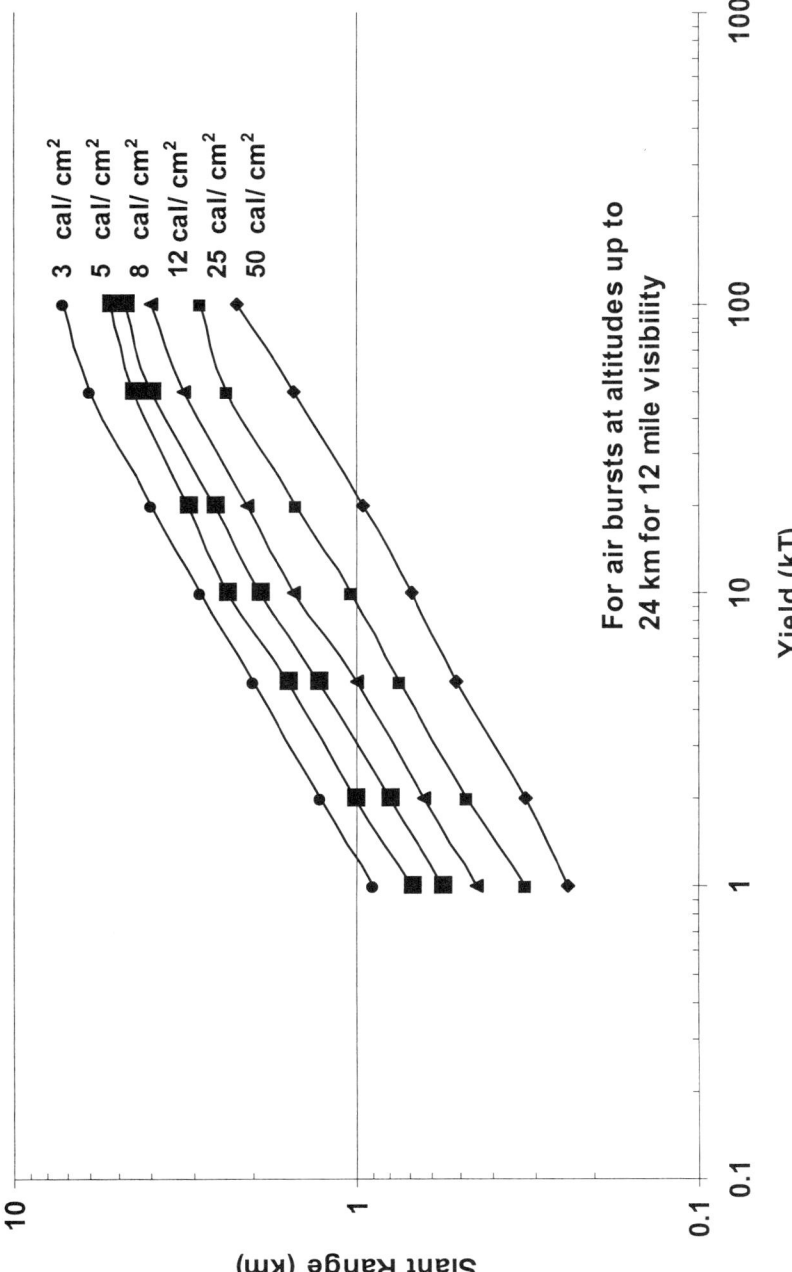

FIGURE 2-4 Slant range for specified radiant exposures and energy yields (Glasstone and Dolan 1977).

(0.65 MeV, 30-keV x-rays) was used (Maruyama and Kuramoto 1987). Details of the exposures are provided by Ichikawa and others (1987). There was some ceramic material surrounding the sample to provide electron equilibrium and some backscatter medium. The in situ exposures from the burst undoubtedly included backscattered gamma rays different from those of the calibration exposures.

Depending on the ambient scattering structures and gamma-ray energy, an increase in the gamma-ray signal due to scattered gamma rays might have been around 20%. The gamma-ray scatter to a single tile from adjacent tiles or structures from the broad-beam exposure from the burst is unknown. This was pointed out by Ichikawa and others (1987), who stated that the actual buildup of the gamma rays in the tiles could have been greater than that in the tiles exposed to ^{60}Co gamma rays without the backscatter of concrete blocks.

Two examples of backscatter are well known. In counting of gamma-emitting samples, a backscatter peak at 0.25 MeV is evident in the gamma-ray spectrum. That results from a 180° scatter of the photon originating from the source (sample) by the surrounding shielding. Thus, backscattered gamma rays contribute to the total signal measured. Similarly, in the calibrating of a pressurized ion chamber with a point source of photons, such as ^{60}Co, scattered photons from the ambient structures or the ground surface contribute to the total signal. To avoid the effect, a shield, usually a lead block, is placed at the source to prevent line-of-sight radiation from the source (a shadow shield). It provides direct measurement of the scattered photon signal The scattered signal is subtracted from the total to obtain the calibration signal.

According to Kaul and Egbert (2000) the DS86 calculations included a component for backscatter and the backscatter signal enhancement is probably less than 5% at 1000 m ground distance.

Quartz Energy Response

All TL materials yield a different—that is increasing—light output with decreasing gamma ray energy. That is due to the higher energy absorption (a larger mass energy coefficient) in the medium with lower energy. Figure 2-5 shows the mass energy-transfer coefficient for SiO_2 (taken for quartz) as a function of energy (Hubbell 1982). Below 0.1 MeV, there is an increase in the coefficient that would increase the energy signal deposited per unit kerma in free air. Maruyama (1983) stated that when the response from ^{60}Co gamma rays was unity, that for 40 keV x-rays was about 3 at a depth of 0.5 cm in the brick.

Figure 2-6 shows the cumulative distributions of fluence at Hiroshima, in the forward and downward directions, as functions of gamma-ray energy estimated in DS86 (Cullings 2000). About 30% of the gamma-ray fluence is below 0.1 MeV, so a signal enhancement of this order of magnitude is possible because of the physical absorption properties of quartz. Figure 2-6 also illustrates the angular dependence of the energy absorption of the brick or tile with respect to orientation to the burst, with about a 20% difference possible. The uncertainty due to orientation was also pointed out by Dean Kaul and colleagues (Roesch 1987).

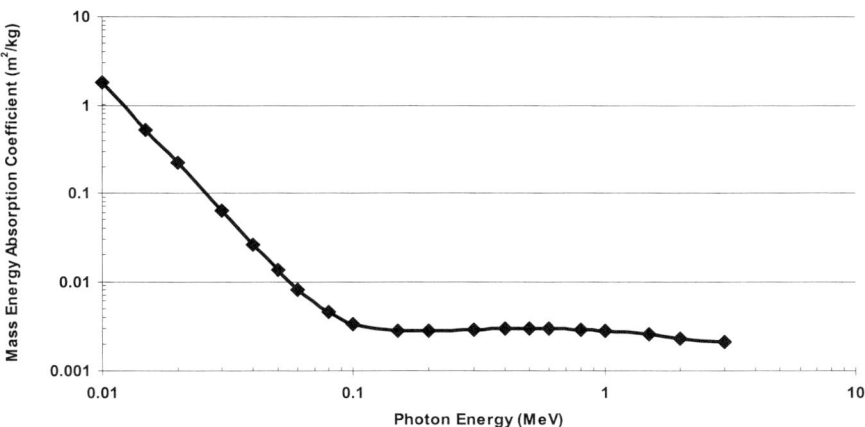

FIGURE 2-5 Photon mass energy-transfer coefficient in SiO₂ (Hubbell 1982).

Figure 2-7 shows the cumulative gamma-ray fluence as a function of energy at Hiroshima with all angles combined for both prompt and delayed gamma rays. De-layed gamma rays contribute most of the exposure and dose from the burst. Figures 2-8 and 2-9 show the cumulative distributions of downward and forward fluence at several ground ranges. They illustrate the slight shift in the energy spectrum—with somewhat more abundant high energy—with increasing ground distance from the

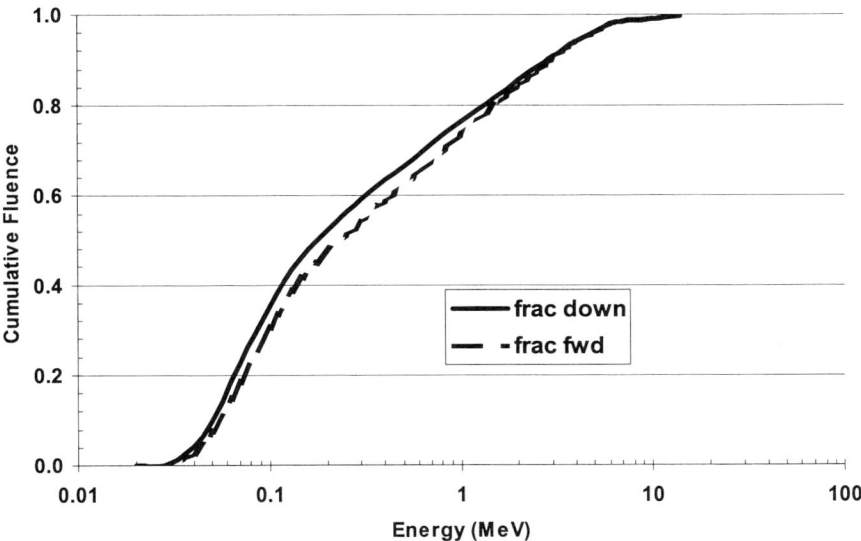

FIGURE 2-6 Hiroshima: cumulative gamma-ray fluence vs. energy at 1000 m downward and forward fluence (Cullings 2000).

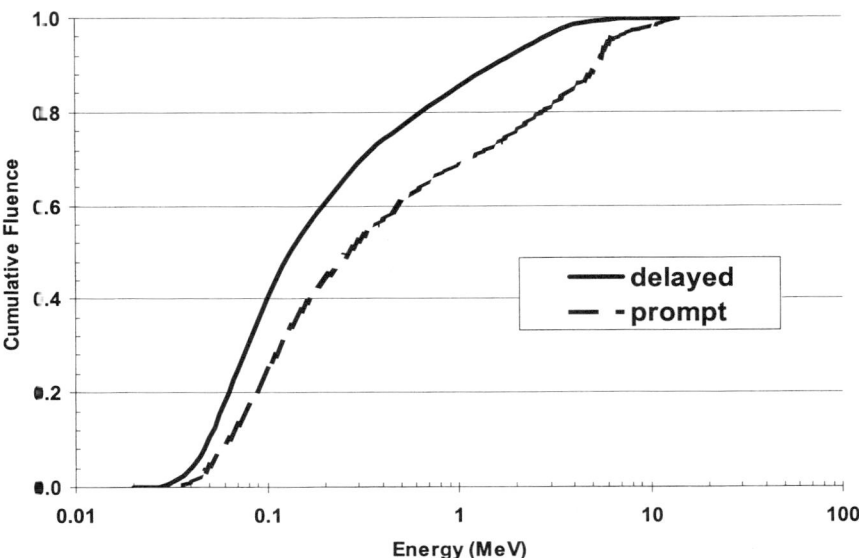

FIGURE 2-7 Hiroshima: cumulative gamma-ray fluence with energy (prompt and delayed) at 1)00 m; all angles combined (Cullings 2000).

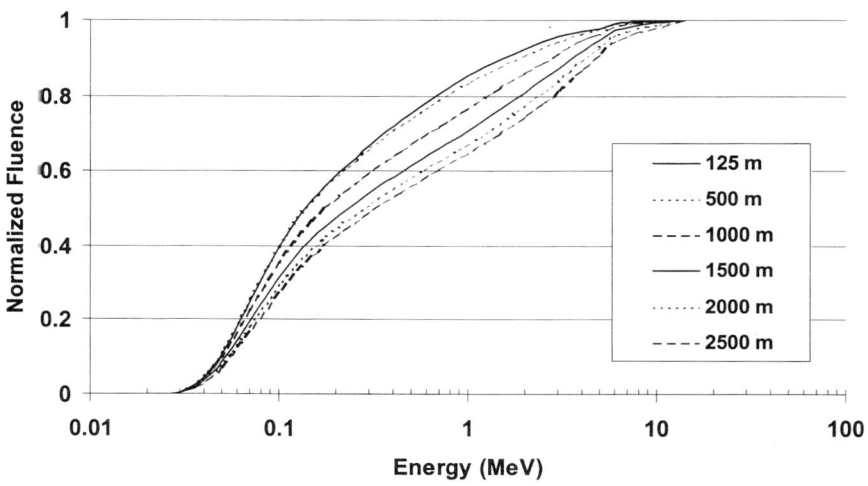

FIGURE 2-8 Hiroshima: cumulative gamma-ray vs. energy by distance (downward fluence) (Cullings 2000).

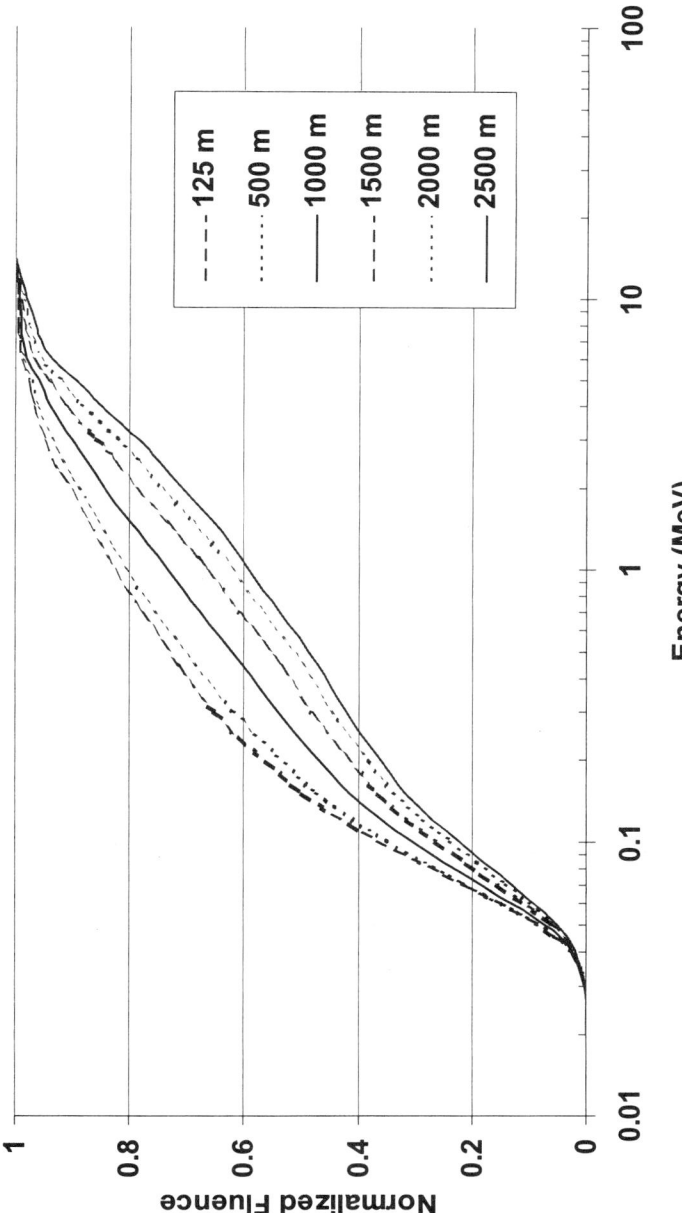

FIGURE 2-9 Hiroshima: cumulative gamma-ray fluence vs. energy by distance (forward fluence) (Cullings 2000).

burst. Again, in Figures 2-6 through 2-9, about 30% of the gamma-ray fluence is below 0.1 MeV, which would lead to a TL signal enhancement.

Background Issues

When a brick or tile is fabricated, the temperature is sufficient to erase or zero the TL radiation signal in the quartz. Later background radiation exposure is stored in the material, with bomb radiation adding to the total TL signal. The background radiation in the brick or tile is due to the uranium and thorium, their decay products, and potassium present in the material itself. External naturally occurring gamma rays and cosmic rays are also part of the background.

The natural radioactivity inclusions in the material emit alpha, beta, and gamma rays. These were assumed to be constant over the lifetime of the sample being discussed here. That is roughly true, but in the initial firing of the material ($\sim$100–1200°C), a fraction of the ^{228}Ra (a decay product in the thorium series) is volatilized, so the inherent background rate would increase over the approximately 40-year postfiring lifetime (Roesch 1987). The inherent background of the brick or tile was determined either by alpha counting or by measuring beta TL. Gamma-ray TL was used to determine the external gamma-ray and cosmic-ray background rates. By placing TL dosimeters on buildings where samples had been taken, the external gamma-ray and cosmic-ray background was measured. The total background exposure subtracted from the samples at Hiroshima and Nagasaki was 0.15–0.32 Gy. Thus, at 2000 m ground range at Hiroshima, for example, the background subtracted is several times that of the bomb signal, and the propagated error in the measurement is 100% (Roesch 1987). Thus background is extremely important especially at the larger distances.

The most important of the other three factors that potentially affect the TL measurements is considered to be the increase in the magnitude (light output) of the TL in response to low gamma ray energy. Although the magnitude of the energy-response correction factor is not known precisely, a downward correction of 20% is plausible on the basis of the above considerations, and was used in the following analyses to derive the sensitivity comparisons. However, as will be shown later, the agreement is better without this correction.

COMPARISON OF TL GAMMA-RAY MEASUREMENTS BETWEEN HIROSHIMA AND NAGASAKI

Figures 2-10 and 2-11 summarize the TL measurements made to date at Hiroshima (Cullings 2000) as corrected to tissue dose (free-in-air [FIA] tissue kerma) for comparison with the DS86 calculations. The data are distinguished by type of reporting unit (roentgen, air dose, tissue dose, or quartz) but corrected in Figure 2-11 to FIA tissue dose so that any differences attributed to reporting units can be identified. None is evident. The DS86 FIA tissue-dose estimates are shown for comparison.

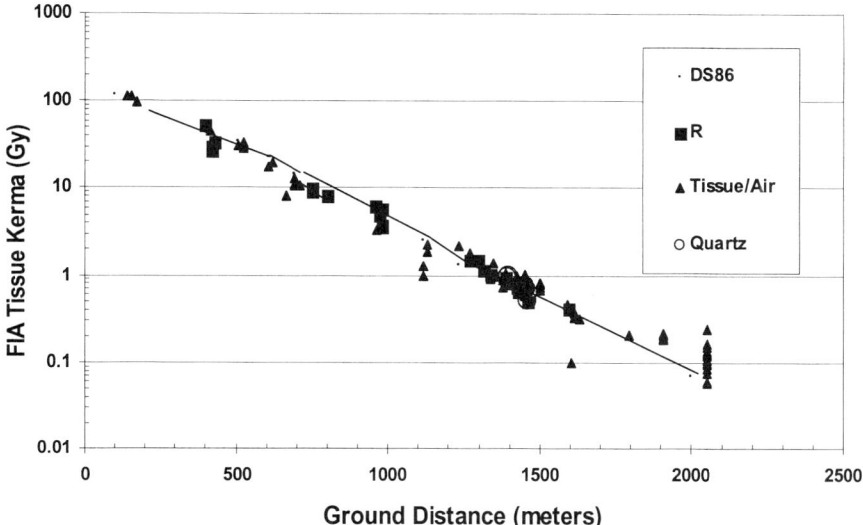

FIGURE 2-10 Hiroshima TL measurements. Original reported units—roentgens, tissue, air, quartz—converted to RA tissue kerma (Cullings 2000).

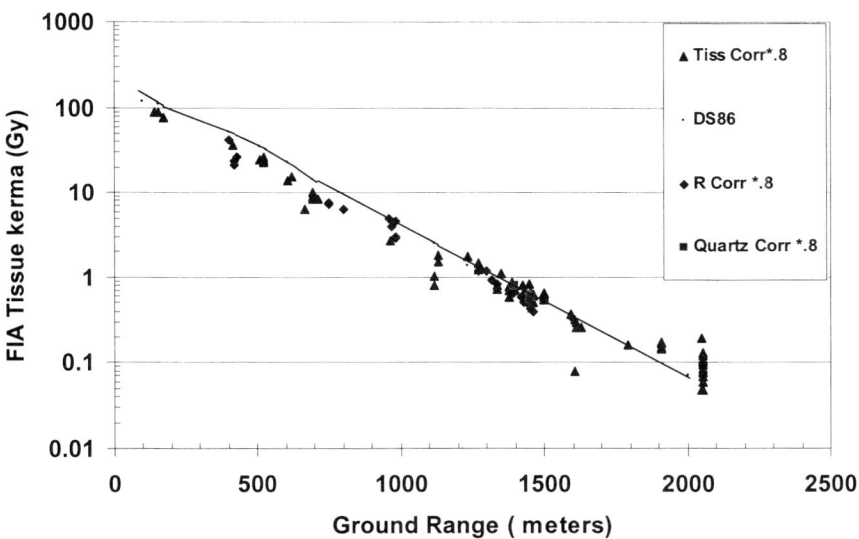

FIGURE 2-11 Hiroshima TL measurements (Corrected). Original reported units—roentgen, tissue, air, quartz—converted to FIA tissue kerma (Cullings 2000).

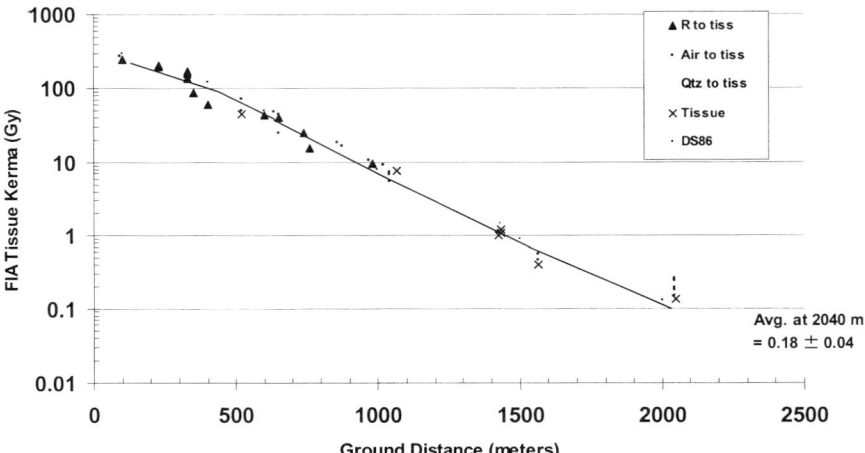

FIGURE 2-12 Nagasaki TL measurements. Original measurements in reported units—roentgens, tissue, air, quartz—converted to FIA tissue kerma (Cullings 2000).

In Hiroshima, there are replicate data at 2050 m. Their arithmetic mean and standard deviation are 0.058 ± 0.089 Gy, as shown in Figure 2-10. The replicate data are important because they demonstrate the acknowledged precision of TL measurements, which, in general, is 20–30%. The mean of the measurements at 2050 m is nearly identical with the DS86 estimate of 0.06 Gy but with large uncertainties (Roesch 1987). Closer to the hypocenter, the agreement with DS86 appears good. Figure 2-11 shows the effect of a 20% correction (downward) to the reported measurements. The result does not appreciably improve the fit of the measurements to DS86.

Figures 2-12 and 2-13 summarize the TL measurements made to date at Nagasaki (Cullings 2000). The data points are differentiated by marker style for the original reported units and corrected to FIA tissue kerma for comparison with DS86. At 2040 m there are replicate measurements, and the arithmetic mean and standard deviation are 0.18 ± 0.04 Gy, compared with the DS86 estimate of 0.11 Gy (Roesch 1987). The same correction factor (0.8) would bring the 2040 m measurements essentially into agreement with DS86. Figure 2-13 shows the same data plotted with a correction factor of 0.8. The fit of the data to DS86 is not improved substantially, and there is a suggestion that a slightly shallower slope for DS86 would provide, in general, a somewhat better fit to the measurements.

One test of the goodness of fit to the DS86 calculations is to examine residuals, that is differences between the data points and the DS86 calculated values. To reduce the scatter of the TL measurements, the measured TL data were averaged in intervals of 100 m. That compresses the number of data points and permits error terms to be calculated directly for the range interval. The interval average data for

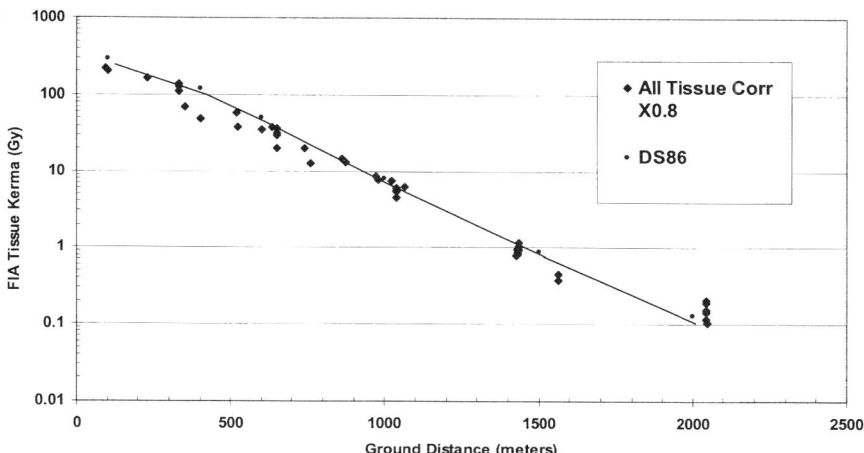

FIGURE 2-13 Nagasaki TL measurements (corrected). Original reported units—roentgens, tissue, air, quartz—converted to FIA tissue kerma (Cullings 2000).

Hiroshima, with error bars, are shown in Figure 2-14. The fractional residual is defined as

residual = (mean of data interval at distance − DS86 value)/DS86[1]

The fractional residual plots for Hiroshima are shown in Figure 2-15. The residual plots show both the original reported data and the data with a 20% (0.8) correction for the assumed energy-response correction. The residual plot in Figure 2-15 indicates that the original data as reported, are in somewhat better agreement with DS86 than the arbitrarily corrected value.

The measured data for Nagasaki were averaged over 100-m distances. The average data, with error bars, are shown in Figure 2-16. The fractional residual plot for Nagasaki is shown in Figure 2-17 (similar to that for Hiroshima) for both the original reported data and the data with a 20% (0.8) correction for the assumed energy response. The residual plot in Figure 2-17 indicates that the original data as reported are in somewhat better agreement with DS86 than the arbitrarily corrected values.

The residuals shown in Figures 2-15 and 2-17 represent the combined effects of errors, both systematic and random, in the DS86 estimates and in the measurements. Roughly stated, the relative magnitude of these combined errors is about 20% for comparatively high doses and about 100% for the natural background component, that is at 2000 m on the ground in Hiroshima. Nevertheless, the test of residuals indicates that because a reduction of 20% does not improve agreement between calculation and measurement, the agreement, fortuitously perhaps, might be to within about ±10%.

---

[1] The DS86 value is interpolated to the midpoint (midway between nearest and farthest measurements) of the distance.

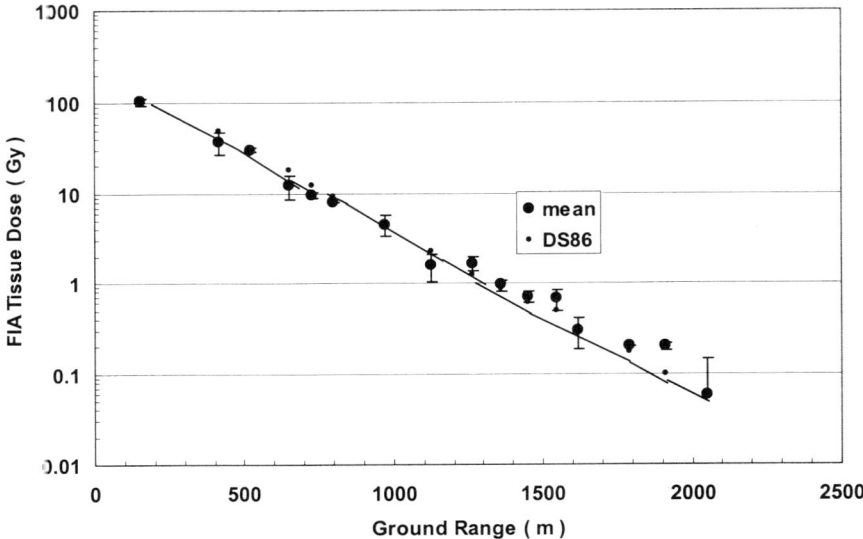

FIGURE 2-14 Hiroshima TL measurements averaged over 100-m intervals.

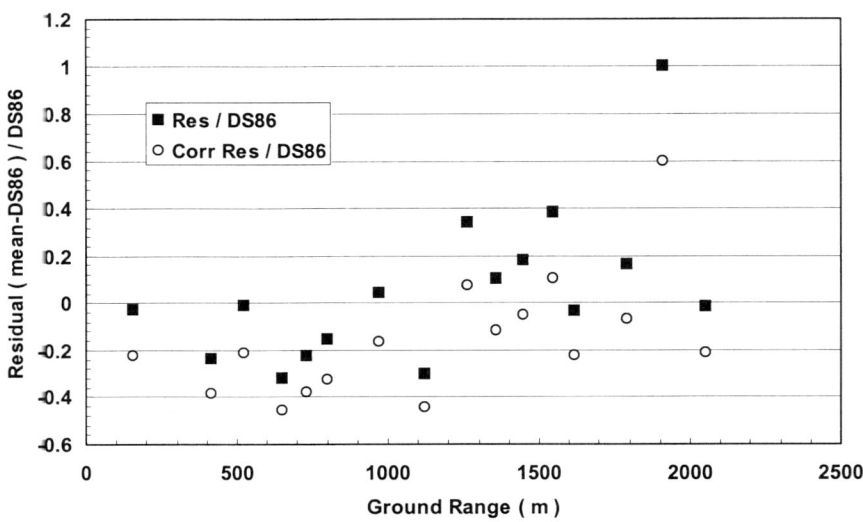

FIGURE 2-15 Hiroshima residuals (mean – DS86)/DS86; original measurement values and values corrected by 0.8.

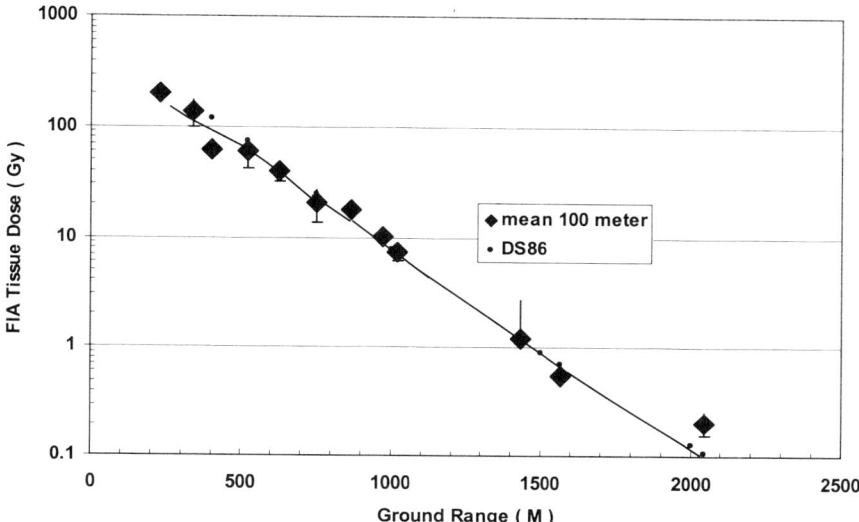

FIGURE 2-16 Nagasaki TL measurements averaged over 100-m intervals.

Figure 2-15 shows that the TLD measurements in Hiroshima appear to be about 20% lower than the DS86 calculation out to 1300 m, then 20% higher at ground distances greater than 1300 m. Nagatomo and others (1995) and Roesch (1987) pointed out this difference for distances greater than 1300 m.

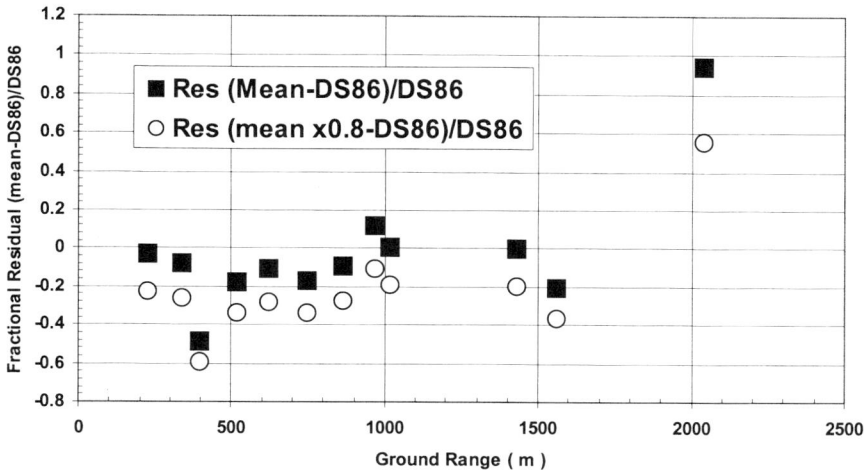

FIGURE 2-17 Nagasaki residuals (mean – DS86)/DS86; original measurement values and values corrected by 0.8.

Figure 2-17 shows that the TLD measurements in Nagasaki are uniformly about 20% lower than the DS86 calculation. The exception at both cities occurs in measurements at greater than 2000 m, where the error in the measurement technique is estimated to be of the order of 100%. Given the overall uncertainty of about 20% for the TLD technique (precision, sample orientation, etc.) there is good agreement of the gamma-ray measurements with DS86 calculations.

Other sources of radiation might provide a small contribution to the TL signal. Neutron activation of short-lived nuclides, such as sodium and calcium, could have added to the TL signal. Fallout from the bomb might have contributed a small amount. There were two tornadoes at Hiroshima within a relatively short time after the burst, and fallout was probably washed away.

If there are actually more fast neutrons at distances in Hiroshima than has been estimated in DS86 they should result in some increase also in the gamma-ray dose at distances because of neutron-capture reactions producing gamma rays.

The four factors inherent in TL measurement uncertainty described cannot be evaluated quantitatively, but the agreement of the TL measurements with DS86 calculations is striking and provides support for the present estimates of gamma-ray dose.

The committee suggests that the working group consider reevaluating the TL measurement data with particular emphasis on whether the energy response of the TL was properly accounted for and on background considerations.

3

Thermal-neutron and Fast-neutron Measurements

Neutrons are central to the operation of the atomic bomb. Fast neutrons are emitted in fission; for every fission, more than a single neutron leaks out of the active material. Measuring those neutrons can indicate the bomb yield directly. However, many fast neutrons can be greatly degraded in energy by materials in the bomb, such as the metal in the casing of the Hiroshima bomb, and by the environment, such as high-explosive gases in the Nagasaki bomb.

DS86 relies on calculations of neutron and gamma-ray fluence and kerma. Although the uncertainty in these calculations can be estimated from the combined uncertainty in the various components of the system, the ability to validate various aspects of the radiation-transport method by comparisons with measurements is essential for demonstrating the overall validity of the method used in DS86. An important comparison with respect to the neutron-fluence calculations at various distances in free air is that between calculated and measured thermal-neutron (low energy) and fast-neutron activation of rocks, building materials, and so on. The neutron dose is smaller than the gamma dose and the neutron kerma (or neutron dose) to survivors results primarily from exposure to higher-energy neutrons. Even thermal-neutron measurements are valuable in testing the transport in air because the thermal-neutron fluence at any site results primarily from downscattering of higher-energy neutrons. Thermal and epithermal neutrons have ranges of only a few meters in air and thus are produced locally. However, the thermal and epithermal fluence incident on a given sample can vary substantially because of variations in local scattering and downscattering. Furthermore, there could be errors in the transport system that affect only the lower-energy neutron transport and not the neutron-dose estimates. Thus, it is desirable to measure the higher-energy fluence directly with activation reactions that have a relatively high-energy threshold.

FAST-NEUTRON ACTIVATION MEASUREMENTS

Following the Hiroshima explosion, Japanese physicists made a number of measurements of the fast-neutron activation of ^{32}S near the epicenter (Roesch 1937). The sulfur was contained in insulation material of electric-power poles. Japanese investigators made these measurements during the first few weeks after the event. The threshold for ^{32}S activation is about 2.5 MeV, and the half-life of the ^{32}P produced is only 14 d. The calculated and measured ^{32}S activation tended to agree well close to the epicenter, particularly when corrections were made for the expected anisotropy due to bomb tilt (Roesch 1987). The exact degree of agreement, however, depends on the assumed yield. In fact, the comparison between ^{32}S activation calculation and measurement was a factor in determining the yield that was used in DS86. The ^{32}S data tended to diverge from DS86 calculations, and the calculated values appeared to be lower than measured by an amount that increased as the distance from the epicenter increased. However, the measured activities at these distances were very low, and the uncertainties very high. Because of the short half-life of ^{32}P, the ^{32}S activation measurements could not be repeated to confirm the original data. Documentation of similar measurements made at Nagasaki has not been found.

THERMAL-NEUTRON ACTIVATION MEASUREMENTS

When DS86 was released, a number of thermal-neutron activation measurements had been made at various slant ranges at Hiroshima and Nagasaki. Additional measurements have since been made of thermal-neutron activation of cobalt (Co) and europium (Eu) and, with a different technique, the generation of ^{36}Cl by thermal neutrons. Those measurements have indicated that the thermal neutrons were more abundant at great distances than was predicted from the neutron spectrum calculated for the bomb explosion by DS86. It appeared that more high-energy neutrons penetrated the iron casing than was calculated. In general, the measured thermal-neutron activation of ^{60}Co at Hiroshima appeared to be higher than the calculated values by an amount that increased as the slant range increased. Although some ^{152}Eu thermal-neutron activation data were also reported, the data for larger distances were too few to confirm the ^{60}Co comparisons. Measurements of ^{60}Co and ^{152}Eu activation at Nagasaki also suggested the possibility of a similar trend, but the data for distances greater than 1000 m was sparse. Near the epicenter at both Hiroshima and Nagasaki, ^{60}Co and ^{152}Eu activation data tended to be about 50% lower than calculated from the DS86 neutron fluence.

RERF has surveyed the literature and communicated with investigators directly and has created a database of all known activation measurements (see Appendix A). Many of the newer measurements were made at increasing distances from the epicenter to resolve the apparent discrepancy observed in the DS86 calculation-measurement comparison.[1] A report including recent ^{36}Cl measurements (Straume and others 1992) confirmed the divergence of DS86 calculations and measurements at great distances in Hiroshima and suggested that the differences could exceed a factor of 10 at slant ranges greater than 1500 m. Newer data for Nagasaki, in conjunction with improved radiation-transport calculations (Kaul and others 1994; Straume and others 1994) appeared to show good agreement between calculation and measurement at all distances, although the data for greater distances were still relatively sparse. However, very recently reported ^{152}Eu activation data for Nagasaki (Shizuma 2000a) again suggest the possibility of similar, although perhaps smaller, discrepancies at great distances. The thermal-neutron activation measurements near the epicenter at Hiroshima continued to be about 50% lower than the calculations.

Since Straume and others reported the apparent discrepancy in thermal-neutron activation calculations at Hiroshima, numerous studies have attempted to explain the lack of agreement. The present committee has examined all the proposed solutions and some new data that have become available since the original publication by Straume and others (1992). The committee and its consultants have examined the various measurement data in great detail to determine whether part or all of the disagreement could be due to measurement errors (including failure to account properly for background contribution) and to determine better the exact extent of the potential discrepancy in calculated DS86 neutron fluence as a function of distance.

Previous assessments have treated all the data as essentially equivalent with respect to accuracy. Because of the varied quality of the reported measurements, the degree of the possible discrepancy might have been overestimated at the greater distances, where many of the data have large uncertainties.

FAST-NEUTRON ACTIVATION MEASUREMENTS OF ^{63}NI

An important new development is the ability to measure the production of ^{63}Ni in copper samples (n,p). Because the activation takes place only at energies above about 1.5 MeV, such measurement provides a method for confirming the ^{32}S measurements at Hiroshima and for directly measuring the high-energy neutron fluence at large distances from the epicenter.

One technique (Straume and others 1996; Ruehm and others 2000b) uses accelerator mass spectrometry (AMS). It is very sensitive and is able to give results on fast-neutron intensity from a few hundred meters to 1500 m.

---

[1] The term discrepancy used in this discussion describes the trend in the measurements in contrast with DS86-calculated values. This is related loosely to the notion that the measurements as a function of slant range, in a semilogarithmic plot, have a shallower slope, corresponding to a greater "relaxation length," than the calculated values. The other aspect of this trend, which is often left implicit, is that the measured and calculated values tend to be equal not near the source, but at some middling slant range of about 800 to 1000 m.

[63]Ni can also be assayed by direct beta-counting (Shibata and others 1994); but because of the long half-life (100 y) and low specific activity, the measurement sensitivity is low except fairly close to the epicenter, where the neutron fluence was high. Preliminary measurements have however been made by this method (Shibata 2000).

[63]Ni measurements should permit an empirical determination of the numbers of fast neutrons emitted by the bomb. Together with the additional thermal-neutron measurements on [36]Cl also being carried out now and a careful reevaluation of the reported thermal-activation data, a complete verification or determination of the neutron spectrum at Hiroshima might become possible, permitting the determination of neutron kerma at the important distance of 1000–1500 m, where the average total doses to survivors lie between 0.2 and 2 Gy. With the new measurement of fast neutrons using [63]Ni, it is hoped that it will be possible to reconstruct a neutron source directly from the fast-neutron measurements combined with the augmented thermal-neutron data.

Thus, the problem in Hiroshima is primarily to explain the neutron discrepancy, assuming that it is real. Previous attempts to provide a source term that might fit the neutron data have led to a neutron source that seemed physically unacceptable (see Chapter 4). The most penetrating fast neutrons in air are above 2.3 MeV. Thus, the basic problem in the neutron measurement is that the long relaxation length in air implied by the thermal-neutron activation data suggests a neutron source with a significantly greater number of neutrons at energies above 2 MeV than expected. But, only neutrons at the same or higher energies cause the capture of neutrons by sulfur and copper. It is exceedingly difficult to construct a neutron source that can provide enough thermal neutrons at 1500 m and not have too large a sulfur capture. It is hoped that the new neutron measurements will allow us to revisit and resolve this discrepancy.

The remainder of this section discusses the committee's evaluation of the various reported activation data and their estimated uncertainty. We estimate the minimum detectable concentrations (MDCs) of the various measurement protocols for different investigators. A number of measurement problems might have resulted in reported activation data that were biased high at low activities. On the basis of this evaluation, we provide our current best estimate of the degree of the calculation-measurement discrepancy relative to distance at Hiroshima and Nagasaki on the basis of best fits of activation measurements to calculated activation as a function of slant range, considering measurement and calculation uncertainties and appropriate background corrections. Finally, we recommend additional measurements that are required to refine the estimate of the discrepancy at Hiroshima.

SUMMARY OF AVAILABLE NEUTRON
ACTIVATION MEASUREMENTS

Thermal-neutron and fast-neutron measurements have been reported on the basis of various reactions. These are summarized in Table 3-1.

TABLE 3-1 Thermal-neutron and Fast-neutron Reactions Used for Measuring Neutron Fluence at Hiroshima and Nagasaki

Reaction	Threshold	T ½	σ^{th} (barns)[a]	σ^{th} (barns)[b]	%[c]
$^{59}Co(n,\gamma)^{60}Co$	Thermal	5.3 y	37	30	27
$^{151}Eu(n,\gamma)^{152}Eu$	Thermal	13.4 y	5900	4400	13
$^{153}Eu(n,\gamma)^{154}Eu$	Thermal	8 y	346	320	41
$^{35}Cl(n,\gamma)^{36}Cl$	Thermal	3×10^5 y	42	28	8
$^{63}Cu(n,p)^{63}Ni$	1.5 MeV	100 y	—	—	—
$^{32}S(n,p)^{32}P$	2.5 MeV	14 d	—	—	—

[a] Thermal capture cross section at 300°K.
[b] Thermal-neutron cross section averaged over energies less than 0.4 eV (Kaul and Egbert 1989).
[c] Fraction of activation from neutrons above thermal, estimated (Kaul and Egbert 1989).

A description of the various samples analyzed, their locations, sample type, degree of shielding, and so on, is included in the database assembled and maintained at RERF. Appendix A describes this database and contains a list of the literature and other sources of the measurements in the database. Additional measurements are still in progress and will be added to the database when available. To evaluate the various data, a questionnaire was sent to each investigator reporting activation data (see Appendix A). The responses to the questionnaire are also included in the database. Letters requesting additional information regarding measurement protocols, uncertainty analyses, and background subtractions were sent to select investigators to clarify various issues. Committee consultants also visited the laboratories of several of the investigators and interviewed the principal investigators.

Not all the investigators responded fully to the questionnaire or the followup requests for clarification. Because the published material did not usually contain sufficient information to evaluate the total uncertainty in reported measurements, the committee and its consultants were required to make their own estimates, as described in Appendix B. Appendix B also discusses the definition and estimates of MDCs for each investigator. For the committee to provide its best assessment of the most critical data sets—those at sites greater than 1000 m—it is essential that investigators be encouraged to provide the necessary information and, as discussed later, agree to cooperate in sharing samples and participating in comparisons.

A number of the measurements described in the RERF database were heavily shielded or were measurements at increasing depth in cores from bridges or buildings. The comparisons between measurement and calculation for these samples involve a shielding correction that was made with the DS86 fluence calculations, so these data have not been used in our assessment of the extent of the disagreement between DS86 calculations and measurements. Any error in the DS86 energy spectrum of incident neutrons would confound the comparison of free-field calculated and measured fluence. Although even the calculation of activation in near-surface samples involves some additional calculation uncertainty, these corrections are (as discussed later) relatively small compared with the apparent discrepancy. They do, however, need to be considered in evaluating the extent of the potential discrepancy in the DS86 neutron-fluence calculations.

Appendix A discusses all the available reported activation measurements that the committee believes were at locations in the direct line of sight of the epicenter and were minimally shielded (that is, near the surface). Preliminary ^{63}Ni data have been provided to the committee and are discussed in this report (Ruehm 2000; Straume 2000a), but they have not been reported and must therefore be considered tentative. Additional preliminary ^{36}Cl data have also been provided to the committee and are discussed in this report (Straume and others 2000), but these data also must be considered tentative and subject to revision based on additional measurements and calculations.

COMPARISON OF ACTIVATION
MEASUREMENTS WITH DS86-BASED CALCULATIONS

Figures 3-1 through 3-4 compare the measured activation of ^{60}Co, ^{152}Eu, ^{36}Cl, and ^{63}Ni at Hiroshima with the corresponding DS86-calculated free-field values. (Where appropriate, a small correction was made to the measured ^{60}Co and ^{152}Eu activation data to correct for shielding; see Appendix B.) To better illustrate the spatial dependence of the data at large distances, both measured and calculated values have been multiplied by the square of the slant range. Assuming that most of the neutrons originate from a point isotropic source at the epicenter, one would expect the thermal fluence, and thus thermal activation, to decrease approximately exponentially with distance, provided that the spectral distribution of neutrons in the epithermal and thermal region remained about the same. The DS86-calculated fluences show this near-exponential decrease, although the relaxation length increases slightly as the distance increases and the neutron spectrum becomes harder (Roesch 1987). The error bars reflect our best estimate of the total uncertainty (1 SD) and, as discussed in Appendix B, include possible errors not included by the investigators in their uncertainty estimates. Thus, our estimates of uncertainty are sometimes larger than those reported by the investigators. As discussed in Appendix B, the total uncertainty includes consideration of possible competing reactions and other sources of error.[2] Also shown in Figures 3-1 through 3-4 is the best fit to the measurement data based on a weighted least-

---

[2] Because we were unable to obtain sufficient information to estimate total uncertainty, the values listed in Table 3-2a and 3-2b are our best estimates and subject to revision. The actual uncertainty is still probably underestimated in as much as several issues discussed later in this chapter and in Appendix B might have contributed to additional errors that cannot be quantified now. As discussed in Appendix B, other reasonable weighting schemes other than that chosen could be used. However, any reasonable weighting scheme would still give little weight to the uncertain measurements at large slant ranges, so the difference between the fit to the measurements and the DS86 calculations would be similar to that shown.

square fit, assuming an exponential decrease with distance (with varying relaxation length). The fitting procedure is discussed in detail in Appendix B. Weighting the data by the inverse of the variance provides a fit that is less influenced by poor-precision data and thus presumably more reflective of the actual spatial trend in the data. Because the ^{63}Ni data are preliminary and subject to revision as additional measurements are completed, no uncertainty estimates are available, and the data have not been fitted. Similarly, the ^{36}Cl data are being reevaluated because of various measurement and calculation issues discussed later in this chapter. Thus, the uncertainties shown reflect only the precision associated with measurements of several aliquots of the same sample.

A cosmic-ray background activation value at the time of measurement has been subtracted from the ^{60}Co and ^{36}Cl measurement data before decaying to "at time of bombing" (ATB) as discussed below and in Appendix C. An appropriate cosmic-ray background for ^{63}Ni is still under investigation, and no correction has been made to the data shown in Figure 3-4.

As can be seen from the figures, notwithstanding the large uncertainties for samples from the more distant sites, the available data clearly indicate that a discrepancy that increases with increasing slant range exists in the DS86 calculations that cannot be explained by measurement error alone. There are only a few samples at very large slant ranges for ^{60}Co, and they all deviate significantly from the best fit to the remaining data if the measurements at the Yokogawa Bridge (see Table 3-2a) are included in the fit. Note that the same investigator analyzed all these samples. They all have large uncertainty, and the uncertainty might be even larger than indicated because of the possible problems regarding cross-contamination and sample-selection bias discussed below. Because of their large uncertainties, these three points have very little influence on the shape of the fitted curve when the Yokogawa Bridge samples are included (see Appendix B). However, the Yokogawa Bridge measurements were shielded, and the effect of the shielding (about 50%) might have been overestimated. It is possible that if the distal data have much greater uncertainty than estimated for this report, the actual discrepancy lies somewhere between the two fitted curves shown in Figure 3-1.

The fit to the ^{152}Eu data (Figure 3-2) suggests a discrepancy similar to that indicated by the fits to the ^{60}Co data. The ratio of measured to calculated, M/C, is about 5–10 at 1500 m. However, all the ^{152}Eu data at the larger slant ranges are from two investigators, and the Shizuma data (indicated by squares in Figure 3-2) (Shizuma 2000a) appear to be consistently higher than data from Nakanishi and others (1983). As shown in Table 3-2a, all the Shizuma data at the larger slant ranges are less than our estimated MDC. However, all the data at great distances might be biased high (and the uncertainty underestimated) because of measurement issues described later in this chapter such as cross contamination of samples, quality control, background activity, cosmic-ray corrections, and other measurement issues.

The more sensitive ^{36}Cl and ^{63}Ni measurements provide the strongest evidence for concluding that the discrepancy is less by a factor of only 5 or less at distances

TABLE 3-2a Hiroshima Neutron Line-of-Sight Measurements at Slant Range over 1000 m (Surface or Near-Surface Samples Except Indicated)

Range, m		Investigator (ref. #)[a]	Site Name	Specific Activity		% S.D. invest.	net activ ATB	% S.D. Rev.	DS86 Free field	M/C	S.D.	stable mg	Estimated MDC, Bq/mg		msmt/ MDC
Ground	Slant			ATB	ATM								ATM	ATB	
^{60}Co (activity in Bq mg^{-1}; estimated cosmic-ray activity = 3.3×10^{-6})															
1014	1168	Shizuma (106)	Hiroshima City Hall	1.05×10^{-1}	1.47×10^{-4}	10	1.03×10^{-1}	12	4.1×10^{-2}	2.5	0.8	36.4	2.23×10^{-5}	1.80×10^{-2}	6.6
1197	1330	Hashizume (23)	Powder Magazine	2.30×10^{-2}	1.72×10^{-3}	19	2.30×10^{-2}	20	6.75×10^{-3b}	3.4	1.2	9.0	1.49×10^{-4}	2.07×10^{-3}	11.5
1295	1419	Kerr (8)	Yokogawa Bridge	5.60×10^{-3}	1.96×10^{-5}	9	4.66×10^{-3}	12	2.2×10^{-3c}	2.1	0.7	675.0	4.34×10^{-6}	1.24×10^{-3}	4.5
1295	1419	Hamada (29)	Yokogawa Bridge	5.15×10^{-3}	1.80×10^{-5}	16	4.21×10^{-3}	20	2.2×10^{-3c}	1.9	0.7	239.0	1.14×10^{-5}	3.26×10^{-3}	1.6
1481	1591	Shizuma (106)	Red Cross Hospital, pipe-1997	2.90×10^{-2}	3.11×10^{-5}	72	2.59×10^{-2}	80	1.20×10^{-3}	22	18	9.6	8.45×10^{-5}	7.88×10^{-2}	0.4
1481	1591	Shizuma (106)	Red Cross Hospital, pipe-1995	1.50×10^{-2}	2.09×10^{-5}	27	1.26×10^{-2}	34	1.20×10^{-3}	11	5	56.4	1.44×10^{-5}	1.03×10^{-2}	1.5
1484	1593	Shizuma (106)	Red Cross Hospital, ladder-1997	3.10×10^{-2}	3.33×10^{-5}	32	2.79×10^{-2}	36	1.20×10^{-3}	23	11	15.0	5.41×10^{-5}	5.04×10^{-2}	0.6
1484	1593	Shizuma (106)	Red Cross Hospital, ladder-1995	3.40×10^{-2}	4.74×10^{-5}	29	3.16×10^{-2}	33	1.20×10^{-3}	26	12	26.3	3.09×10^{-5}	2.21×10^{-2}	1.5
1703	1799	Shizuma (106)	Hiroshima Bank of Credit-1997	1.10×10^{-2}	1.18×10^{-5}	82	7.92×10^{-3}	109	2.20×10^{-4}	36	41	15.3	5.32×10^{-5}	4.96×10^{-2}	0.2

1703	1799	Shizuma (106)	Hiroshima Bank of Credit-1995	2.10×10^{-2}	2.93×10^{-5}	14	1.86×10^{-2}	18	2.20×10^{-4}	85	30	73.0	1.11×10^{-5}	7.96×10^{-3}	2.6
4571	4608	Shizuma (106)	Army Food Storehouse	0.00	0.00							2.2	3.70×10^{-4}	2.65×10^{-1}	0.0
^{152}Eu (activity in Bq mg^{-1}; estimated cosmic-ray activity = 8×10^{-5})															
849	1028	Shizuma (18)	Choukaku-ji	2.4	2.28×10^{-1}	17	2.40	20	1.45	1.7	0.6	0.036	1.16×10^{-1}	1.220	2.0
880	1054	Nakanishi (11)	Tamino's House (Kawara-machi)	3.5	5.54×10^{-1}	31	3.50	32	1.120	3.1	1.4	0.033	4.27×10^{-1}	2.700	1.3
881	1055	Shizuma (18)	Hiroshima Prefectural Office	2.2	2.09×10^{-1}	9	2.20	12	1.110	2.0	0.6	0.069	6.08×10^{-2}	6.40×10^{-1}	3.4
893	1065	Shizuma (18)	Honkei-ji	1.5	1.42×10^{-1}	27	1.50	29	1.010	1.5	0.6	0.031	1.36×10^{-1}	1.430	1.0
912	1081	Shizuma (18)	Enryu-ji	2.4	2.28×10^{-1}	63	2.40	64	8.70×10^{-1}	2.8	2.0	0.027	1.58×10^{-1}	1.660	1.4
924	1091	Shizuma (18)	Yorozuyo Bridge stone wall	1.4	1.33×10^{-1}	21	1.40	22	7.90×10^{-1}	1.8	0.7	0.098	4.28×10^{-2}	4.50×10^{-1}	3.1
927	1093	Shizuma (18)	Shingyo-ji	2	1.90×10^{-1}	45	2.0	47	7.70×10^{-1}	2.6	1.4	0.038	1.10×10^{-1}	1.160	1.7
949	1112	Shizuma (18)	Teramachi stone wall	1.7	1.61×10^{-1}	53	1.70	54	6.46×10^{-1}	2.6	1.6	0.046	9.07×10^{-2}	9.56×10^{-1}	1.8
988	1146	Shizuma (18)	Hiroshima Radio Station	1.8	1.71×10^{-1}	22	1.80	24	4.71×10^{-1}	3.8	1.5	0.049	8.61×10^{-2}	9.07×10^{-1}	2.0
1017	1171	Shizuma (18)	Hiroshima City Hall	1.1	1.04×10^{-1}	27	1.10	29	3.73×10^{-1}	2.9	1.2	0.043	9.65×10^{-2}	1.020	1.1
1060	1208	Nakanishi (10)	Hiroshima City Hall	1.15	1.21×10^{-1}	13	1.150	15	2.65×10^{-1}	4.3	1.5				
1060	1208	Nakanishi (10)	Hiroshima City Hall	1.02	1.07×10^{-1}	50	1.020	51	2.65×10^{-1}	3.8	2.3				
1060	1208	Nakanishi (10)	Hiroshima City Hall	1.15	1.21×10^{-1}	13	1.150	16	2.65×10^{-1}	4.3	1.5				
1163	1300	Shizuma (18)	Kozen-ji	1.1	1.04×10^{-1}	45	1.10	47	1.15×10^{-1}	9.6	5.3	0.022	1.90×10^{-1}	2.000	0.5
1197	1330	Shizuma (18)	Iwamiya-cho	0.9	8.54×10^{-2}	50	8.99×10^{-1}	52	8.80×10^{-2}	10	6.1	0.034	1.25×10^{-1}	1.310	0.7
1255	1383	Nakanishi (10)	Hiroshima University	0.53	6.84×10^{-2}	45	5.29×10^{-1}	46	5.50×10^{-2}	9.6	5.3				
1255	1383	Nakanishi (10)	Hiroshima University	0.18	2.32×10^{-2}	72	1.79×10^{-1}	73	5.50×10^{-2}	3.3	2.6				

(continued)

TABLE 3-2a Hiroshima Neutron Line-of-Sight Measurements at Slant Range over 1000 m (Surface or Near-Surface Samples Except Indicated) (*Continued*)

Range, m Ground	Slant	Investigator (ref. #)[a]	Site Name	Specific Activity ATB	ATM	% S.D. invest.	net activ ATB	% S.D. Rev.	DS86 Free field	M/C	S.D.	stable mg	Estimated MDC, Bq/mg ATM	ATB	msmt/ MDC
1274	1400	Nakanishi (11)	Hiroshima University	0.41	4.31×10^{-2}	44	4.09×10^{-1}	45	4.80×10^{-2}	8.5	4.6				
1274	1400	Nakanishi (11)	Hiroshima University	0.36	3.79×10^{-2}	39	3.59×10^{-1}	40	4.80×10^{-2}	7.5	3.7				
1274	1400	Nakanishi (11)	Hiroshima University	0.48	5.05×10^{-2}	50	4.79×10^{-1}	51	4.80×10^{-2}	10	5.9				
1298	1422	Nakanishi (11)	Hiroshima University	0.34	3.57×10^{-2}	35	3.39×10^{-1}	36	4.00×10^{-2}	8.5	4.0				
1328	1449	Nakanishi (11)	Hiroshima University	0.114	1.20×10^{-2}	32	1.13×10^{-1}	34	3.23×10^{-2}	3.5	1.6				
1335	1456	Shizuma (18)	Hiroshima University, Primary School	0.5	4.75×10^{-2}	64	4.99×10^{-1}	65	2.97×10^{-2}	17	12	0.028	1.52×10^{-1}	1.60	0.3
1357	1476	Shizuma (18)	Kyo Bridge, Railing	0.55	5.22×10^{-2}	71	5.49×10^{-1}	72	2.48×10^{-2}	22	17	0.047	9.00×10^{-2}	9.48×10^{-1}	0.6
1370	1488	Shizuma (18)	Teishin Hospital (Communications Hospital)	0.51	4.84×10^{-2}	43	5.09×10^{-1}	44	2.25×10^{-2}	23	12	0.053	7.98×10^{-2}	8.41×10^{-1}	0.6

^{36}Cl (activity in ^{36}Cl atoms / 10^{15} Cl atom; estimated cosmic-ray background = 108)

Range, m Ground	Slant	Investigator (ref. #)[a]	Site Name	Specific Activity ATB	ATM	% S.D. invest.	net activ ATB	% S.D. Rev.	DS86 Free field	M/C	S.D.
889	1045	Ruehm et al. (182)	Sinkoji gravestone	2500	2.50×10^{3}	15	2392		2600	0.9	0.3
1029	1181	Ruehm et al. (182)	Ganjioji-gravestone	400	4.00×10^{2}	10	292		680	0.4	0.1
1061	1209	Straume	Hiroshima	700	7.00×10^{2}	17	592		474	1.2	0.4

1140	1279	City Hall	et al. (181)	300	3.00×10²	27	192	92	2.1	0.8
1217	1348	Tokneiji-gravestone	Ruehm et al. (182)	240	2.40×10²	16	132	150	0.9	0.3
1225	1355	Jyunkyoji-gravestone	Ruehm et al. (182)	260	2.60×10²	29	152	140	1.1	0.5
1386	1502	Hosenji-gravestone	Ruehm et al. (182)	300	3.00×10²	53	192	39	4.9	3.0
1470	1580	Teishin Hospital	Straume et al. (181)	212	2.12×10²	36	104	20	5.2	2.4
1470	1580	Red Cross Hosp.	Straume et al. (181)	325	3.25×10²	59	217	20	11.0	7.0
1606	1708	Red Cross Hosp.	Straume et al. (181)	198	1.98×10²	39	90	8	11.0	5.4
1606	1708	Postal Savings	Straume et al. (181)	185	1.85×10²	47	77	8	10.0	5.6
		Postal Savings	Straume et al. (181)							

^{63}Ni (activity in atoms/micro gram Ni; no cosmic-ray background subtracted)

948	1111	Soy Sauce	Straume, Marchetti, Ruehm, et al. (181)	0.51	0.60	0.50	1
1014	1168	City Hall	"	0.41	0.41	0.36	1
1304	1427	Univ. elem. School	"	0.17	0.17	0.055	3
1461	1572	Univ. radio-isotope bldg	"	0.15	0.15	0.028	5

[a] ^{36}Cl and ^{63}Ni data are all preliminary and subject to change; see text. DS86 values are free-field except where indicated.
[b] The numbers in parentheses refer to Table 2 in Appendix 1.
[c] Shielding factor of 0.75 applied.
[d] Shielding factor of 0.51 applied.

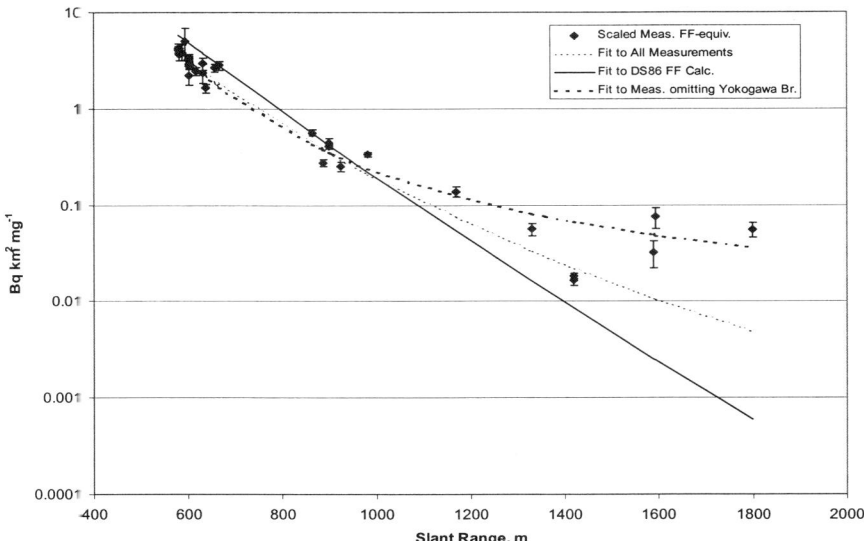

FIGURE 3-1 Hiroshima ⁶⁰Co measurements (scaled) vs. DS86 free-field. Including high precision but heavily shielded Yokagawa Bridge samples (see Table 3-2a) has large effect on fit. The error bars are 1 SD.

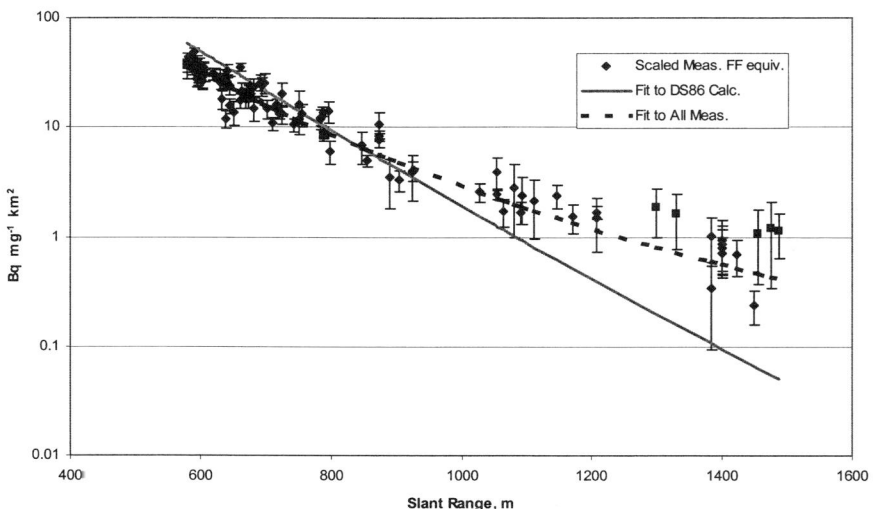

FIGURE 3-2 Hiroshima ¹⁵²Eu scaled DS86 free-field and measured FF-equivalent values. Data shown as square symbols are less than the estimated MDC (see Table 3-2a). The error bars are 1 SD.

less than 1500 m. However, the ^{63}Ni data are preliminary. Furthermore, measurement problems are associated with the reported ^{36}Cl data, discussed below, and must still be resolved. Until the ^{36}Cl data have been reevaluated and an uncertainty analysis of the data performed, the ^{36}Cl results must be considered incomplete. Thus, we have not attempted to fit the data shown in Figure 3-3, as was done for the ^{60}Co and ^{152}Eu data. Figure 3-3 includes some new preliminary results of ^{36}Cl activation in granite gravestones performed by Ruehm (2000a). They suggest reasonably good agreement with DS86 at distances out to about 1300 m. However, the data have not been corrected to free-field values. The DS86 values are free-field estimates, so the difference between measured and calculated activation is probably larger than shown in the figure.

Only a few preliminary ^{63}Ni measurements are available, but they suggest (see Figure 3-4) that the discrepancy for higher-energy neutrons is smaller than that for thermal neutrons; this perhaps reflects systematic uncertainty in the calculation of thermal activation at the more distant sites. Clearly, additional high-quality measurements are needed to confirm the preliminary data.

As can be seen in Figures 3-1 through 3-4, the thermal-neutron data continue to disagree with the DS86 calculations near the epicenter (and thus also with the ^{32}S data) at Hiroshima. The calculated results are about 50% higher than the measurements in all cases. Thus, it is important to perform additional ^{63}Ni measurements on surface samples near the epicenter to validate the ^{32}S measurements. It is impor-

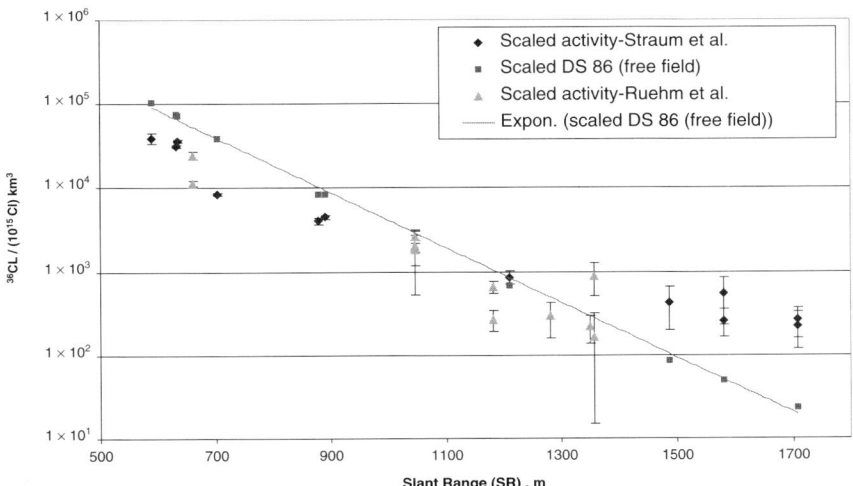

FIGURE 3-3 Hiroshima ^{36}Cl data vs. DS86 free-field calculations. All data are preliminary and subject to revision. No corrections have been made to account for differences between free-field activation and in the in situ sample (Ruehm and others 2000a; Straume 2000a). The error bars are 1 SD.

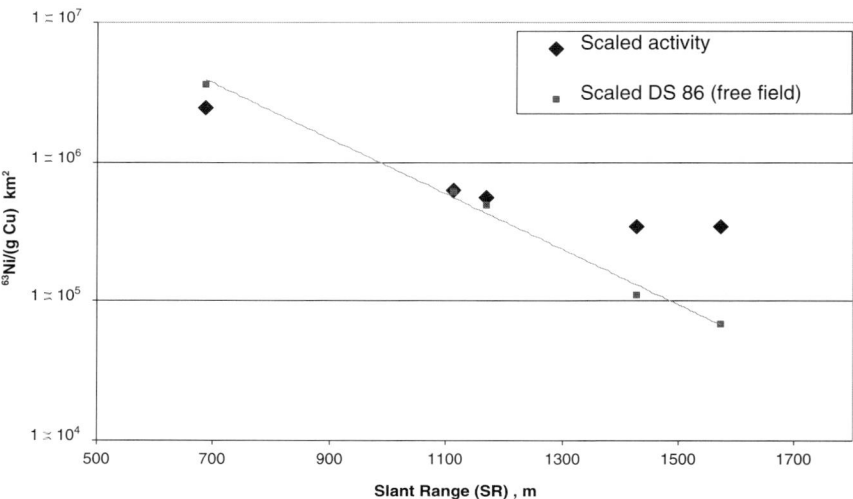

FIGURE 3-4 Preliminary [63]Ni data vs. DS86 free-field calculated values. No cosmic-ray background subtraction has been made and data have not been corrected to free-field (Straume 2000). Solid line is exponential fit to free-field DS86 calculations.

tant to note, however, that the [32]S data are higher than the DS86 calculations according to the newest cross-section sets and improved transport calculations (see Chapter 4), whereas the europium, chlorine, cobalt, and nickel data are lower. If the [63]Ni data do not confirm the [32]S data, the disagreement near the epicenter might reflect errors in height of burst, yield, source term, and radiation transport that are unrelated or only partially related to the discrepancy at large distances.

Table 3-2a compares the reported activities for all sites at slant ranges greater than 1000 m. The uncertainty reported by the investigators often does not account for the uncertainty in the measurement of stable-element concentration or the calibration factors that apply to counting efficiency and the assay of stable-element concentration. Thus, Table 3-2a lists the investigator-reported uncertainty, as well as revised estimates that account for additional sources of possible error. Appendix B describes how the uncertainties were calculated. Note that the uncertainty estimates for the [36]Cl data are based only on the precision estimated from multiple analyses and must be considered tentative. Table 3-2a also lists estimates of the MDCs of ^{60}Co and [152]Eu obtainable by particular investigator methods for comparison with the reported values.

Initial measurements of Hiroshima copper samples show promise that it will be possible to obtain good measurements of bomb-neutron fluence as a function of distance out to substantial distances at Hiroshima (1500 m and beyond). Present measurement values, so far unpublished, are too preliminary for adequate evaluation in this report.

Figure 3-5 combines all the data for Hiroshima at slant ranges greater than 1000 m in a plot of M/C vs. slant range with approximate measurement errors and a ± 30% assumed calculation uncertainty (1 SD). The plotted error bars represent the 95% confidence limits (2 SD). Data less than the estimated MDC in Table 3-2a are not included, because the committee believes that data below the estimated MDC are unreliable.

Figures 3-6 and 3-7 compare ^{60}Co and [152]Eu activation measurements with DS86 at Nagasaki. Included are recently reported [152]Eu measurements by Okumura and Shimazaki (1997) and by Shizuma (2000a). The Shizuma data suggest that the DS86 calculation discrepancy also exists at Nagasaki. The MDCs of these measurements are also shown in Table 3-2b. Again, the data at distal locations, particularly the Okumura data, are sufficiently close to the MDC to be suspect. The new Shizuma data appear to be systematically higher than those of Nakanishi (see Figure 3-7 and Table 3-2b), just as the Shizuma [152]Eu data at Hiroshima appeared to be higher than those of other investigators. The slope of the DS86 results appears to decrease somewhat

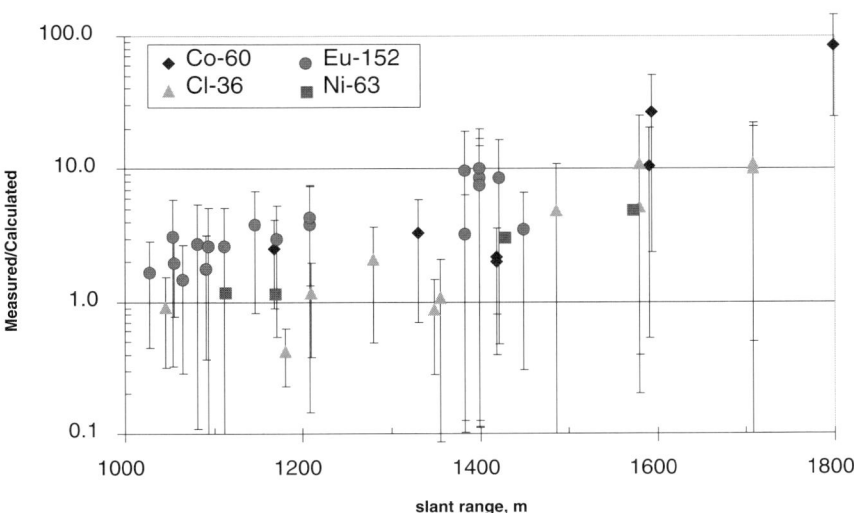

FIGURE 3-5 Ratio of measured to calculated activation. Data from Table 3-2a. Error bars are 2SD. Data below MDC are not included. No uncertainty estimates are available for preliminary [63]Ni data.

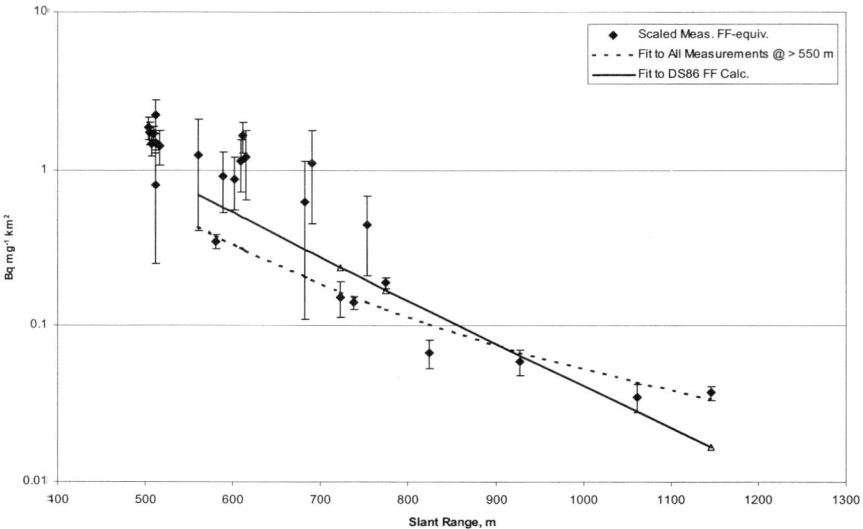

FIGURE 3-6 Nagasaki ^{60}Co. The error bars are 1 SD.

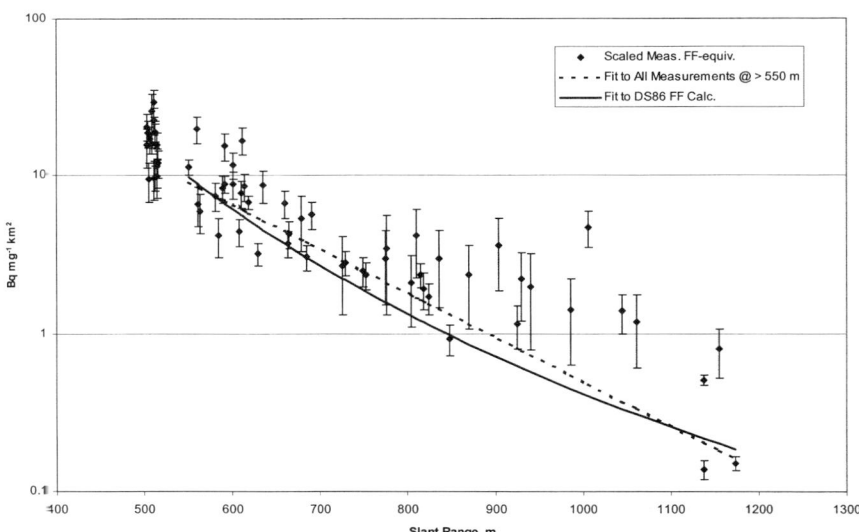

FIGURE 3-7 Nagasaki ^{152}Eu. The error bars are 1 SD.

TABLE 3-2b Nagasaki Neutron Line-of-Sight Measurements at Slant Range over 1000 m (Surface or Near-Surface Samples)[a]

| Range, m | | Investigator | Site Name | Specific Activity | | % S.D. invest. | net activ ATB | % S.D. Rev. | DS86 | M/C | S.D. | stable mg | Estimated MDC, Bq/mg | | msmt/ MDC |
Ground	Slant			ATB	ATM								ATM	ATB	
60Co (activity in Bq mg⁻¹; estimated cosmic-ray activity = 3.3×10⁻⁶)															
935	1062	Shizuma (19)	Mitsubishi Steel	0.033	4.6×10^{-5}	18	0.031	21	0.02	1.5	0.5	30.5	2.67×10^{-5}	1.91×10^{-2}	1.7
1030	1146	Hashizume (23)	Comm School	0.024	1.78×10^{-3}	8	0.0244	10	1.10×10^{-2}	2.2	0.7	9	1.50×10^{-4}	2.10×10^{-3}	6.6
152Eu (activity in Bq mg⁻¹; estimated cosmic-ray activity = 8×10⁻⁵)															
871	1006	Nakanishi (25)		4.66	0.325303	25	4.66	25	0.39	12	4.7	0.023	0.613	3.87	1.2
916	1045	Shizuma (177)		1.27	0.088656	27	1.27	28	0.29	4.4	1.8				
934	1061	Shizuma (177)		1.05	7.33×10^{-2}	48	1.05	49	0.26	4.0	2.3				
1020	1137	Nakanishi (171)	Anakoboji Temple	0.446	3.63×10^{-2}	19	0.446	20	0.15	3.0	1.1	1.91	7.38×10^{-3}	0.10	4.4
1020	1137	Nakanishi (171)	Anakoboji Temple	0.288	2.34×10^{-2}	17	0.288	18	0.15	1.9	0.7	1.16	1.22×10^{-2}	0.17	1.7
1020	1137	Nakanishi (171)	Anakoboji Temple	0.08	6.51×10^{-3}	20	0.08	21	0.15	0.5	0.2	1.17	1.21×10^{-2}	0.16	0.5
1020	1137	Nakanishi (171)	Anakoboji Temple	0.094	7.65×10^{-3}	36	0.094	37	0.15	0.6	0.3	1.72	8.20×10^{-3}	0.11	0.8
1020	1137.28	Iimoto (164)	Anakoboji Temple	0.491	3.61×10^{-2}	8.1	0.491	9.8	0.15	3.3	1.0	1.91	9.37×10^{-3}	0.13	3.7
1020	1137.28	Iimoto (164)	Anakoboji Temple	0.336	2.73×10^{-2}	9.5	0.336	11	0.15	2.2	0.7	1.16	1.54×10^{-2}	0.22	1.5

(continued)

TABLE 3-2b Nagasaki Neutron Line-of-Sight Measurements at Slant Range over 1000 m (Surface or Near-Surface Samples)[a] (Continued)

| Range, m | | Investigator | Site Name | Specific Activity | | % S.D. invest. | net activ ATB | % S.D. Rev. | DS86 | M/C | S.D. | stable mg | Estimated MDC, Bq/mg | | msmt/ MDC |
Ground	Slant			ATB	ATM								ATM	ATB	
1039	1154.35	Shizuma (177)	Sakamoto-cho	0.6	4.88×10^{-2}	33	0.6	34	0.135	4.4	2.0	0.06	7.22×10^{-2}	0.93	0.6
1060	1173.29	Nakanishi (171)	Ide residence	0.113	9.20×10^{-3}	19	0.113	19	0.12	0.9	0.3	1.56	$9.04 \times 10^{-}$	0.12	0.9
1060	1173.29	Nakanishi (171)	Ide residence	0.11	8.95×10^{-3}	11	0.11	12	0.12	0.9	0.3	1.51	9.34×10^{-3}	0.13	0.9
^{36}Cl (activity in ^{36}Cl atoms per 10^{15} Cl atoms; estimated cosmic-ray background = 122)															
1090	1187	Straume et al. (181)	Mitsubishi Steel	340	340	6	218		164	1.3	0.4				
1160	1261	"	Fuchi Middle School	210	210	16	88		80	1.1	0.4				

[a] ^{36}Cl and ^{63}Ni data are all preliminary and subject to change; see text. DS86 values are free-field except where indicated.
[b] The numbers in parentheses refer to Table A-2 in Appendix A.
[c] Shielding factor of 0.75 applied.
[d] Shielding factor of 0.51 applied.

more than that for Hiroshima at large slant ranges in Figures 3-6 and 3-7. That probably reflects the different sources for the two cities. A much larger fraction of the thermal-neutron fluence (and thus activation) at Nagasaki was due to delayed neutrons from the fireball than at Hiroshima (about 33% vs. 8%) (Roesch 1987).

Few ^{36}Cl data have been reported at low activities in Nagasaki (see Table 3-2b). The plot of M/C at distances greater than 1000 m for Nagasaki (Figure 3-8) suggests that when uncertainty in both measurement and calculation is considered, the available reliable data do not support the existence of a discrepancy with distance. However, good data at large distances are sparse, and there are no good data at distances beyond a 1300-m slant range. Because the neutron fluence at Nagasaki was much lower than that at Hiroshima, activation at 1000–1200 m at Nagasaki corresponds to that at about 1050–1350 m at Hiroshima. If the lack of a discrepancy up to 1300 m at Nagasaki is confirmed by additional data, it would strongly suggest that the discrepancy at Hiroshima at low activities cannot be due only to measurement error and background subtraction errors in that these errors would also have been expected to occur at comparable activities at Nagasaki. Thus, it is very important to analyze additional samples from distances beyond 1000 m at Nagasaki. Because the source spectrum of neutrons at Nagasaki is very different from that

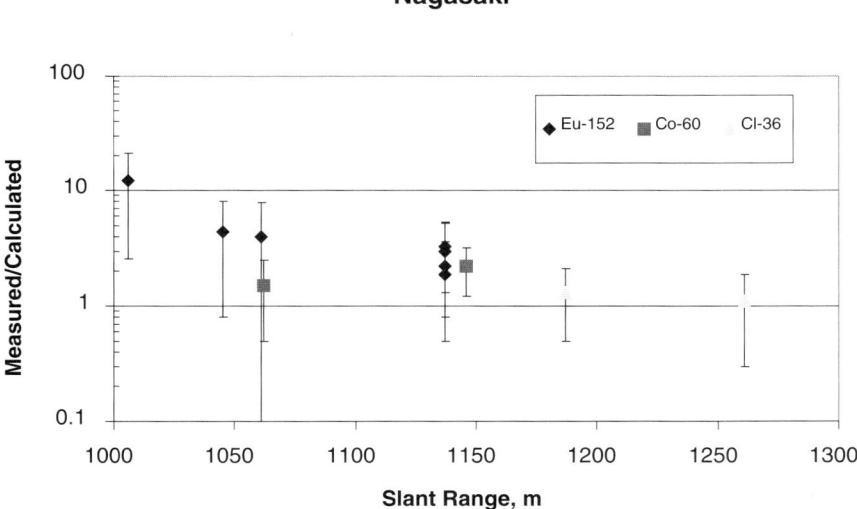

FIGURE 3-8 Ratio of measured to calculated activation. Data from Table 3-2b. Error bars are 2SD. Data below MDC are not included.

at Hiroshima and over one-third of the activation at large distances is calculated to be due to the delayed-neutron component, as opposed to only about 8% at Hiroshima, a similar discrepancy at Nagasaki might also indicate that at least some of the discrepancy is due to terrain or other effects, rather than source-term errors. The terrain at Nagasaki is irregular compared with Hiroshima, so one might expect increased scattering and possibly more degraded spectra at large distances (and correspondingly greater thermal activation) than calculated with DS86.

MEASUREMENT ISSUES

A more detailed discussion of measurement and uncertainty issues is in Appendix B, which is a rigorous evaluation by special scientists to the committee of the uncertainty in reported measurement data. The following summarizes some of the discussion in Appendix B.

Cross Contamination of Samples

Because the half-lives of ^{152}Eu and ^{60}Co are relatively short and it has been over 50 years since the events, the activation measurements reported in recent years, which include most of the data at larger slant ranges, are very low. The amounts of natural europium and cobalt in the samples are small, so enough must be extracted from the sample (that is the sample must be enriched sufficiently) to produce adequate measurement sensitivity. Many of the reported ^{152}Eu and ^{60}Co measurements at the greater distances are only slightly higher than or even less than the estimated MDCs (see Table 3-2). Thus, those data are highly suspect and should be used with caution. Because the lower-activity samples are so close to the MDCs, even very slight cross contamination of low-activity samples by previously prepared high-activity samples might have resulted in overestimation of the lowest-activity samples and underestimation of total uncertainty. As discussed below and in Appendix B, it appears that strict quality-control procedures to prevent cross contamination (use of blanks, blind sample analysis, and so on) were not generally followed; thus, one cannot discount the possibility that cross-contamination of samples occurred, particularly when the range of sample activity often varied over several orders of magnitude. For example, the range of ^{60}Co activity in the samples prepared by Shizuma (1998) varied over a range of more than 1000. Even slight cross-contamination could have severely contaminated the lowest-activity samples. All the analyses of low-activity samples rely on chemical procedures to enrich the cobalt, europium, chlorine, and nickel and are thus subject to cross-contamination. Cobalt samples were prepared by using a milling machine to scrape chips from the surface of steel samples for chemical separation (Shizuma and others 1998). Tools were used to grind concrete and rock samples to prepare samples for ^{36}Cl analysis. Although the committee has no direct knowledge that any such contamination occurred, the apparent flattening out of the measured activities for europium and cobalt at great distances and the fact that some samples were apparently obtained at similar distances that did not show measurable activity (see Appendix B) suggest that the possibility exists and should be explored further. The ^{152}Eu and ^{60}Co data at great distances in particular are also suspect because of possible selection bias in deciding which samples were actually exposed to bomb neutrons. There is some evidence that some samples were assumed unexposed if spectrum photo peaks were not clearly evident and thus data that might have reduced the average activity at distal sites was not reported (see Appendix B).

Quality Control

It appears that few investigators performed rigorous quality control or participated in a quality-assessment program. Provenance of samples is not always well documented, and the mislabeling of samples is always a potential problem.

Data-quality assessment has been defined by the US Environmental Protection Agency (EPA) as a procedure by which existing data can be used to make a value judgment or decision (US EPA 1994a; US EPA 1994b). The US Department of Energy has established a similar approach for environmental data based on the guidelines published by EPA (Tindal 2000). The procedure is used to establish whether a data set is adequate for making some decision or estimate. It can be used to answer two fundamental questions:

- Can a decision or estimate be made with a desired level of confidence, given the quality of the data set?
- How well can the data set be expected to perform over a wide range of possible outcomes?

Assessment of environmental data must unambiguously establish the reliability of the data. Furthermore, monitoring data must be confirmed before application of statistical evaluation or other interpretations. Uncertainties arise from sampling procedures (sampling variance) and analytical procedures (analytical variance). The process of systematic and independent verification of data and the associated uncertainties or variances is data-quality assessment (Miller and Fitzgerald 1991).

In the case of environmental measurements to ascertain the neutron yield of the Hiroshima and Nagasaki bombs, sampling variance is minimized by considering only samples whose exact locations at the time of the bombs are well established. Sample shielding must also be known. Sample variance also depends on sample size (Kratochvil and others 1984) and concentration of the analyte in the sample (Boyer and others 1985). For some measurements, analyte concentrations are quite small, and large quantities of environmental materials are needed. In other cases, heterogeneity in the distribution of the analyte in the sample dictates that a specific sample size be used. Those kinds of variance can be assessed with methods described by Clark and others (1996).

Analytical variance reflects a combination of systematic and random errors associated with measurement. Systematic errors, which can be detected and corrected, can be reduced by various means, including analysis of standard reference materials, analysis of blank samples containing no analyte, use of a different method to obtain the same result, and interlaboratory comparisons. In interlaboratory comparisons, several laboratories analyze identical samples with the same or different methods.

Ideally, analytical variance is accurately reflected in the estimate of uncertainty reported with an analytical result. This variance reflects the total uncertainty arising from chemical and physical manipulations of the sample during preparation for analysis and from the established uncertainty of the analytical method used. It is always desirable to test any analytical result obtained with one method by repeating the measurement with a different method. Good agreement between measurements made with multiple methods affords confidence in the result and is useful in establishing the uncertainty in the measured value.

Apparently, only a few intercomparisons were carried out to test counting accuracy; to our knowledge, no intercomparisons were carried out to test sample preparation, and so on. Preparation and analysis of duplicate aliquots were rare, as was splitting of samples. Exchange of prepared samples between laboratories doing the same types of analyses (such as germanium gamma-spectrometric counting) to test counting accuracy was apparently not routine. That is unfortunate, because the MDCs for a given type of analysis varied considerably from investigator to investigator, and an intercomparison of low-activity samples might have identified a systematic bias. For the ^{152}Eu samples, at least, the half-life is long enough so that available samples could still be shared and counted in facilities with lower background. It would also be desirable to analyze some of the same concrete samples for both ^{36}Cl and ^{152}Eu at the more distant sites.

Background Activity

An important potential source of error in the reported activity at the time of bombing for sites at very large slant ranges is the correction for environmental background activity. All investigators corrected for background counting errors from their detector system due to radiation, but some of the corrections can be highly uncertain, particularly at low activities. Most investigators did not correct for activation of the samples in situ by neutrons produced by cosmic-ray secondaries in the atmosphere. Appendix C discusses cosmic-ray activation in some detail. Shizuma (1999) attempted to calculate the contribution of cosmic rays or measure the activation in laboratory reagents. He concluded that the contribution was negligible. For the case of ^{60}Co, he apparently based that conclusion on a comparison with his measured ^{60}Co activation in reagents, which is somewhat uncertain and only about one-fifth of the activation measured by another investigator (see Appendix C). It is important to note that the estimated cosmic-ray activation must be compared with the measured activation, not with the activation at the time of the bombing after correction for decay. Because ^{60}Co activation measured in the 1990s has decayed by as many as 10 half-lives, the possible cosmic-ray activation contribution at the time of measurement is not a negligible fraction of the measured activity in samples obtained at great distances from the epicenter and could account for about 20–30%, or more, of the reported activity at the time of the bombing at the most distant locations, as shown in Table 3-2.

Cosmic-ray Corrections

The cosmic-ray corrections shown in Table 3-2 are our current best estimates based on measurements in laboratory reagents and calculations (see Appendix B and Appendix C). The cosmic-ray activation of samples in situ could be higher or lower than that in laboratory reagents that were presumably stored in a building and thus could have been substantially shielded from the full cosmic-ray fluence, so the actual ^{60}Co background for some samples might have been even larger than our crude estimate. The ^{152}Eu cosmic-ray activity based on the reagent measurements and calculations appears to be too small to account for any significant error in the measurements, particularly because the decay correction due to the elapsed time between the bombing and sample measurement is only up to about a factor of 10 compared with about a factor of 1000 for ^{60}Co. Owing to the short half-lives of ^{60}Co and ^{152}Eu, all the cosmic-ray activation took place in situ after the bombing and saturation (production rate = decay rate) should have been essentially achieved.[3]

Measurements by Straume (2000b) of the attenuation in deep concrete and at sites far from the epicenter indicate that the cosmic-ray contribution is significant for measurements beyond about 1200 m—^{36}Cl/Cl of around 100×10^{-15}. The cosmic-ray activation in concrete can be expected to vary somewhat depending on the history of the material (sand, and so on) used to make the concrete, so the estimated background subtraction for ^{36}Cl might not be the same for all samples.[4] Cosmic-ray ^{36}Cl activities from less than 100×10^{-15} to 600×10^{-15} have been reported for sands and rocks (see Appendix C). As discussed in Appendix C, because of the lower geomagnetic latitude of Japan, a cosmic-ray fluence of about one-half to three-fourths that at higher latitudes is likely, therefore the value estimated from

---

[3] An additional consideration with respect to ^{152}Eu in roof and wall tiles is whether the cosmic-ray activation of tiles that were blown off buildings and thus exposed to cosmic rays while lying on the ground or elsewhere was greater than would have occurred if the tiles remained in place.

[4] Note that because of the long half-life of ^{36}Cl (300,000 y), all the activation took place over many thousands of years before the production of the concrete in the sample. For much of that time, the sand and other material used to make the concrete was probably shielded to some degree from cosmic rays. The fact that the estimated activities are only a few percent of that expected if complete saturation (production rate = decay rate) of a surface source had occurred substantiates this.

the core measurements is reasonably consistent with the reported data. However, values 2–3 times higher than estimated from the core measurements, depending on the history of the sample materials, would also not be inconsistent with the reported data. In Table 3-2a and 3-2b, we have used the mean of 108×10^{-15} for Hiroshima and 22×10^{-15} for Nagasaki with an estimated coefficient of variation (CV) of 25%.

Because it is clear that cosmic-ray activation could have been an important contributor to the measured ^{60}Co and ^{36}Cl activity for samples at great distances and because the actual cosmic-ray contribution depends on the location and exposure conditions, it is important to obtain good measurements of cosmic-ray activation for samples similar to those analyzed (that is steel plates or lightning rods for ^{60}Co and rocks, tiles, and concrete for ^{36}Cl) far enough from the epicenter for bomb-fluence effects to be neglected so that the cosmic-ray exposure conditions can be considered comparable. As discussed in Appendix C, cosmic-ray activation in thick slabs of material first increases because of spallation of the much higher incident-energy spectrum, which produces a shower of evaporation neutrons, and then decreases exponentially with depth, so estimating the cosmic component by measuring the variation in activity in cores, as was done for ^{36}Cl, would not be an appropriate substitute for measurements of surface samples at great distances.

Other Measurement Issues

Several other measurement issues considered in Appendix B might increase the uncertainty reported by various investigators. The location of some samples was not always documented. Such samples as roof tiles were blown off buildings and later recovered, so their location at the time of the bombing is certain only to within several tens of meters. On the basis of a calculated DS86 free-field relaxation length of about 126 m from thermal activation at slant ranges of 500–1000 m, an error of 10 m in the slant range would result in an error of about 8% in the calculated activation.

In many cases, the surface of the buildings from which concrete cores were obtained for ^{36}Cl analysis apparently were subjected to repairs, so the depth at the time of bombing corresponding to a particular measured value is quite uncertain. Some surface samples, in particular those for ^{36}Cl analysis, might also have been contaminated because of dilution of the surface ^{36}Cl activity by chlorine in rainwater or enhancement of the signal due to repairs made with contaminated cement. Furthermore, ^{36}Cl can be produced by activation of ^{39}K, and this could have resulted in an additional error in the ^{36}Cl/total chlorine estimates for samples with high potassium and relatively low chlorine content (Straume 2000b).

The AMS technique used to measure ^{36}Cl results in a higher sensitivity for ^{36}Cl activation than for europium and cobalt. However, a detailed uncertainty analysis has not yet been reported for these data, and replicate measurements of the same sample indicate fairly poor measurement precision even at high activity (Straume 2000b). The problems of surface contamination and potassium activation discussed above are also of concern. Furthermore, thermal activation in concrete is very sensitive to water content and elemental composition of the concrete and surrounding medium. As discussed below, that might result in a substantial difference between the DS86 free-field calculation and the actual activation in a sample core from a building or bridge. Similarly, the activation in samples taken from granite gravestones might not be adequately reflected by the DS86 free-field calculations at ground level. Thus, the comparisons between the [36]Cl surface activity and DS86-calculated values are more uncertain than previously reported and are being re-evaluated by the various authors.

The [63]Ni data, also from AMS measurement, have relatively poor sensitivity at great distances, particularly when the amount of stable nickel in the copper sample is high. If the concentration of nickel in the sample is relatively high, the [63]Ni created by thermal activation can also be high. Before the [63]Ni data can be considered reliable, there must be a complete uncertainty analysis that accounts for potential errors due to the chemical separation of nickel from copper, AMS measurement, thermal-neutron activation of nickel, and uncertainty in total nickel content of the sample.

UNCERTAINTY IN CALCULATED ACTIVATION

It is important to note that the calculation of activation even in a near-surface line-of-sight sample using free-field DS86 fluences is somewhat uncertain. That would be true even if the calculated DS86 fluence in air were highly accurate. Lillie and others (1988) estimated the uncertainty in the free-field fluence at about 20% (1 SD). The calculated activations shown in Figures 3-1 through 3-8 are based on free-field values 1 m above the ground. Sample-specific calculations were carried out for only a few samples—generally samples that required a substantial shielding correction. The energy spectrum traversing the sample was assumed to be the same as that in the air. The actual spectrum—particularly the energy spectrum for the neutron energies below about 0.1 keV—traversing the sample, even near the surface and due to backscattering, depends heavily on the sample material, water content, sample geometry, and depth. The slowing down and the resulting spectral distribution of these neutrons and the thermal neutrons they produce in the sample account for most of the observed activation even in near-surface samples (Roesch 1987). The incident thermal neutrons account for only a small fraction of the activation. For most samples, the effect of sample depth, orientation, composition etc. (in particular water and boron content), was estimated on the basis of benchmark calculations. Kaul and Egbert (2000) presented the results of some benchmark calculations illustrating the sensitivity of the calculated fluence to building height, sample orientation, depth of sample, and so on. Straume and others (1994) also discussed the results of benchmark calculations and the resulting uncertainty in the activation calculations. Although all the estimated corrections from free-field to surface samples are relatively small (around tens of percent at most) and none of these effects alone can explain the large apparent discrepancy between measured and calculated values at great distances, the total uncertainty in the calculated activation due to all perturbations might be large enough (50–100%) to obscure the magnitude of the possible discrepancy in the DS86 free-field air fluence, especially at greater distances.

Furthermore, the true calculation of the activation in a sample depends on a folding together of the activation cross section with the energy spectrum of neutrons traversing the sample. In DS86, only a few energy groups were used to describe this energy distribution. In fact, DS86 contains only one thermal group consisting of all neutrons below 0.4 eV, although more extensive calculations have since been made (see Chapter 4). The activation by thermal neutrons is thus estimated by assuming the spectral shape in this bin to be a Maxwellian distribution with average energy corresponding to a temperature of 300°K and calculating a weighted thermal cross section (Kaul and Egbert 1989). Because the epithermal cross sections have substantial structure, particularly for ^{60}Co, the calculations must assume a spectral shape in the epithermal energy bands (generally 1/E if an appropriately weighted cross section is to be used). For a sample imbedded even slightly, the shape of the spectrum in the epithermal and thermal region will depend on the material temperature and depth. The thermal activation will come primarily from a slowing of incident epithermal neutrons in the sample.[5] For the case of ^{60}Co, about 25% of the activation is due to neutrons above thermal energy from resonances in the cross section (see Table 3-1). Thus, the uncertainty in the calculated activation is sensitive to the actual energy distribution of the fluence in the first few centimeters of the sample and to the change in this distribution from that calculated for the free field because of slowing in the sample itself.

The DS86 neutron fluence varies substantially with height above the ground (over 40% higher thermal fluence at 1 m than 25 m at 1500 m). That reflects the greater thermalization and backscattering of neutrons by soil than by air as shown in Table 3-3.

The thermal fluence near the ground is larger than at 25 m, but the epithermal fluence is lower (Cullings 2000). The calculated activation assumes that the thermal and epithermal fluences near the surface of a structure are similar to those 1 m above the ground, on the basis of limited benchmark calculations. However, variations of around 20–30% can easily occur, depending on the location of the sample and its composition (water and boron content in particular). At the more distant slant ranges, where the line-of-sight angle is fairly small, it is also possible that the low-energy fluence is greatly underestimated by DS86 because of increased scattering

---

[5] Many of the investigators reporting on thermal-neutron activation measurements in cobalt or europium estimated the DS86 (calculated) activation by multiplying the published (DS86) fluence by the thermal-activation cross section at 300°K shown in Table 3-1 rather than by a Maxwellian-weighted average as SAIC apparently does (Kaul and Egbert 1989). The difference is about 20–30% for thermal activation.

TABLE 3-3 DS86-calculated Neutron Fluence (n/cm^2) at 1 m Versus 25 m (Cullings 2000)

Ground Range	Thermal Neutrons			Epithermal Neutrons[a]		
	Height 1 m	Height 25 m	Ratio	Height 1 m	Height 25 m	Ratio
1000 m	2.6×10^{10}	2.0×10^{10}	1.3	3.8×10^9	5.4×10^9	0.70
1500 m	5.7×10^8	4.1×10^8	1.4	8.6×10^7	1.2×10^8	0.72

[a] 29–100 eV energy bin.

of neutrons by terrain features and structures. Finally, the activation cross sections themselves have uncertainty, and some, such as the ^{63}Ni cross sections, are very uncertain (Egbert 1999).

Thus, calculation uncertainty might account for some of the scatter in the calculated M/C ratios of samples collected at about the same distance as shown in Figures 3-5 and 3-8 and some of the observed discrepancy. Furthermore, previous comparisons of M/C ratios as a function of slant range, using a nonweighted regression that failed to reduce the influence of the higher uncertainty in more-distant measurement data and to include the possible calculation uncertainty, might have unduly emphasized the distant data and led to an overestimate of the discrepancy in the DS86 neutron fluence at large slant ranges.

SUMMARY, CONCLUSIONS, AND RECOMMENDATIONS

A number of measurement issues might explain in part why the thermal-activation measurements reported for large slant ranges are too high. They include possible cross-contamination, sample-selection bias, giving too much weight to data below the MDC, and inadequate background subtraction. However, all the Hiroshima activation measurements still show a consistent pattern of discrepancy, with measured values exceeding calculated values at greater distances, but lower than calculated values near the epicenter. When corrections are made to account for cosmic-ray activation and samples with very high uncertainty (near or less than the MDC) are disregarded or given less weight, the discrepancy appears to be somewhat smaller than previously reported. The preliminary ^{63}Ni data suggest that the M/C ratio at 1500 m might be only around 3–5 for higher-energy neutrons.

The ^{152}Eu data, although of questionable accuracy, still tend to support a larger discrepancy than the ^{60}Co activation data, so it would be useful if some of the available samples could be reanalyzed by multiple laboratories in an intercomparison exercise. That might indicate possible measurement bias at low activities and explain the larger apparent bias. Participation in this exercise with extremely low-background counting rooms would be very helpful. Such intercomparisons would require the cooperation of the owners of the processed and previously counted samples.

Some of the discrepancy (about 30% higher calculated neutron kerma at 1300 m) is already known to be due to the DS86 method on the basis of changes in cross sections and improvements in transport methods discussed in Chapter 4. Some additional discrepancy may be explained by errors in the calculation of activation. However, on the basis of the discussion earlier in this chapter, calculation uncertainty can probably account for no more than about a factor of 2 of the M/C discrepancy for any particular sample.

Neutron fluences at Nagasaki at 1000–1300 m correspond to fluences at Hiroshima at about 1200–1500 m. Thus, the apparent good agreement at Nagasaki at distances up to about 1300 m implies that the discrepancy at Hiroshima cannot be due primarily to measurement errors or an error in the cosmic-ray background subtraction. However, there are few data at distances beyond 1000 m at Nagasaki. It is therefore important to obtain additional data at Nagasaki at slant ranges of 1000–1400 m to confirm the agreement with DS86 calculations.

The nearly exponential decrease in both calculated and measured scaled activation (see Figures 3-1 through 3-4) at Hiroshima suggests that the remaining discrepancy at large distances can be explained by a slightly harder source spectrum that would allow more higher-energy neutrons to penetrate to greater distances and thus produce a larger local thermal and epithermal fluence at that distance. However, a higher proportion of higher-energy neutrons from either the bomb or the fireball would make the current agreement with the ^{32}S measurements worse to a greater extent than it would improve the agreement at large distances (Kaul and others 1994).

The committee offers the following recommendations:

• The highest priority should be given to making additional measurements of ^{63}Ni at sites near the Hiroshima epicenter to compare with the ^{32}S measurements and to making as many measurements as possible at distances greater than 1200 m.

• Additional ^{36}Cl measurements should be obtained. Because of concerns about the reliability of ^{36}Cl measurements from concrete, priority should be given to measurements in granite unless investigators can provide a protocol for measurements in concrete that address the issue of reliability. Because the ^{36}Cl data are the only thermal-activation data with sufficient sensitivity to provide reliable results at large distances, it is important to resolve the surface-contamination and potassium-activation issues for ^{36}Cl and to obtain more reliable estimates of the uncertainty in these data, particularly at large distances.

• Measurements of ^{63}Ni at Nagasaki are needed to determine whether a similar but smaller discrepancy exists at activities comparable with those corresponding to Hiroshima slant ranges of 1200–1500 m. Emphasis should be on measuring line-of-sight minimally shielded samples. Additional ^{36}Cl measurements are also desirable at these distances in Nagasaki.

• Because the actual cosmic-ray activities cannot be accurately calculated, owing to the dependence on location and local scattering and attenuation, it is important that background measurements be made on samples similar to those already analyzed but collected further away at locations and in positions similar to those of the actual bomb samples. That is particularly important for [63]Ni, for which no cosmic-ray background data are available,[6] and would be highly desirable for [152]Eu.

• Shibata (2000) has demonstrated the capability of measuring at least the close-in copper samples for [63]Ni with direct liquid-scintillation beta counting. Duplicate aliquots of samples collected close to the epicenter should be prepared and analyzed with both AMS and beta counting techniques.

• The possibility that some of the reported activity in the samples collected at large slant ranges is a result of cross-contamination during sample preparation and chemical-enrichment or measurement-selection bias should be further investigated. Investigators should be asked to document that procedures used to ensure that such cross-contamination did not occur and to see whether blanks and quality-control samples were a part of every batch sample preparation and analysis. All new measurement programs should include a quality-assurance component with established data-quality objectives and procedures to provide assurance that cross-contamination problems will be identified and eliminated.

• For the committee to provide its best assessment of the most critical data sets, those at sites greater than 1000 m, it is essential that the investigators that have reported activation measurements be encouraged to provide the necessary information and agree to cooperate in sharing samples and participating in intercomparisons.

• Substantial environmental data on isotopes produced by neutron activation in the Hiroshima and Nagasaki bombs exist and are documented in the RERF database (Appendix B, Table B-1). Whether those data will be useful in resolving the apparent discrepancy between measured and calculated neutron fluences depends on the completion of a data-quality assessment for this database. The working group should establish data-quality objectives and a data-quality assessment procedure to evaluate the existing data, with procedures described by EPA (US EPA 1994a) and DOE (Tindal 2000). This activity must be a joint and integrated effort involving both US and Japanese researchers and involving researchers who make the measurements and theoreticians who estimate neutron fluences.

• The uncertainty in the DS86 activation calculations has not been thoroughly investigated. A thorough uncertainty assessment of DS86 or its successor should include an assessment of the uncertainty in these calculations. The committee further recommends that any new dosimetry system utilize the best current technology and cross sections in calculating the neutron activation of samples. Furthermore, when measurements and calculations are compared, the method used for the calculations should be clearly specified.

---

[6] Although calculations indicate that the cosmic-ray activation in copper is likely to be small relative to the measured signal (Ruehm and others 2000a) even at large distances, the cosmic-ray fluence might be increased considerably in high-Z materials because of spallation of incident very-high-energy neutrons, and so the calculated fluence based on cosmic-ray spectra in air might be too low.

4

Radiation Transport Calculations

Even when DS86 was accepted and implemented, it was acknowledged that thermal-neutron fluence as estimated from cobalt-activation measurements "contradicts the calculated values by an ever-increasing factor that is five at 1000 m" (Roesch 1987). That led the contributors of the neutron measurements in the final DS86 report to say the following (Roesch 1987):

> If the measured cobalt activations were accepted as correct representations of thermal fluences and the assumption then made that the calculated fluences on the ground are low by a factor that applies to all energies, then the proportion of neutron kerma in the mixed radiation field beyond 1000 m at Hiroshima would change from insignificant to significant. This leaves the possibility, however unlikely in our collective expert judgment, that the calculated neutron-kerma values are wrong. No known evidence contradicts this hypothesis. Therefore, the conclusion of this chapter on neutron measurements must be that the neutron doses are in doubt until further work is done.

The unsatisfactory performance of DS86 in calculating thermal-neutron activation was also noted in an independent review of the new dosimetry system performed by the National Research Council (NRC 1987).

Despite those problems, DS86 was implemented because of the improvements it offered over the previous dosimetry system, particularly in gamma-dose agreement with TLD measurement and in organ-dose calculation. At the time, those advantages and all the technical advantages of DS86 outweighed the inexplicable discrepancies in some of the thermal-neutron data. DS86 agreed with the limited data on sulfur activation above the threshold of about 2.5 MeV and did not show any important discrepancy with calculations for thermal-neutron activation of europium. No one knew the magnitude or importance of discrepancies in the cal-culation of thermal neutrons. Up to the point where they became important as ac-tivation markers of the output of the bomb, thermal neutrons were not considered important and were collapsed into a single energy bin in the neutron cross-section used for DS86. Furthermore, the gamma component of dose was considered to dominate the total absorbed dose to the organs of the survivors. Given those cir-cumstances and the obvious advantages of DS86 as an overall dosimetry system for RERF, it was implemented despite unresolved misgivings about thermal neu-trons. The radiation-transport calculation technology on which DS86 is based had advanced substantially since the preceding dosimetry system (T65D) was com-pleted, and the technology has continued to advance since DS86.

As more advances were made in the technology enabling radiation-transport calculations, as more-sophisticated activation-measurement techniques have been developed, and as more measurements have been made, the discrepancy between the neutrons measured in materials still retaining isotopic markers of the bomb neu-trons and the neutron fluence calculated with DS86 has become more widely ap-preciated. However, recognition of a discrepancy between activation measurement of in situ materials and the activation calculated with DS86 does not identify the cause of the discrepancy. Physicists who have dealt with this type of problem have had and still have differences of opinion about the source of the discrepancy. Given the complexity of the calculations and measurements involved and the inherent un-certainties surrounding the bombings, any of several factors, singly or in combi-nation, could be the source of the problem.

The RERF dosimetry system, as already noted, is actually a series of very complex components consisting of, first, the calculated output spectrum of the bomb; the calculated portion of the detonation spectrum that actually escapes from the bomb casing; the transport of that spectrum through the fireball created by the explosion; the calculation of the interactions and geometric distribution of the ra-diations in the air in the city; and finally the transport of the air-over-ground spec-trum through whatever shielding exists. As the spectrum of neutrons released from the nuclear explosion traverses those various environments, it is constantly chang-ing because of the interaction of the neutrons with the elements in their path. The calculation of those interactions depends not only on the elements encountered be-tween the nuclear explosion and the point of measurement, but also on the neutron cross sections of those materials, the abundance of the materials in the path of the neutrons, and how well the neutron cross sections of the materials are known. All those complex nuclear dimensions depend on the number and energy spectrum of the neutrons generated by the fissile material in the bomb.

The bombs detonated in Hiroshima and Nagasaki were different from each other in the materials they contained and how they were designed. Therefore, the neutron spectra that were generated and that escaped from those two bombs were different. The Hiroshima bomb was unique in that it was the only one of that exact design ever detonated. In contrast, bombs of the Nagasaki-type have been built and tested numerous times in numerous countries. Thus, the leakage spectrum, yield, benchmark measurements, and calculation codes for the Nagasaki bomb are much better known and understood than those for the Hiroshima bomb.

When neutrons interact with elements in materials, they lose energy and the process generates gamma rays and lower-energy neutrons. At the point of nuclear explosion, neutrons can have several million electron volts of energy. By the time they create thermal-neutron activation, the neutron energy is measured in electron volts. The final dose to a person and the production of isotopes by thermal neutrons are determined not only by the neutrons released in the explosion and their transport through air, but also by their interactions with terrain and structures that act as shielding, which further dissipate neutron energy and reduce the quality and magnitude of the radiation exposure at any point of interest. To ensure accuracy in the doses calculated for survivors, all the terms in the calculation, from detonation to dose, must be understood and modeled accurately.

The radiation doses for the survivors in Hiroshima and Nagasaki are not known *a priori*. Just as in any other radiation exposure event, the doses for irradiations of individuals have to be reconstructed. In Hiroshima and Nagasaki the doses for survivors are calculated by DS86. As pointed out in Chapter 2, the gamma-ray component of the organ doses to individuals constitutes the major portion of the total absorbed dose and is in good agreement with direct measurements of gamma-ray signals left by the bombs in the quartz grains in sample cores from exposed bricks and roof tiles. The smaller neutron doses are less certain and are more difficult to verify. Many efforts have been made to confirm calculated neutron doses by measuring the radioactivity induced in elemental material present at the time of the bombing. Benchmark measurements of this type have been made repeatedly for atomic-bomb tests, in radiation-accident reconstruction, and in occupational-radiation monitoring programs. Although the principles of such measurements are well known, their conduct depends totally on finding radioisotopes of sufficient activity to be measurable more than 50 years after the event. The constraints imposed by time and the destruction by the bombing itself make the absolute accuracy of activation measurements difficult at best. Thus, the ability to confirm any radiation-dose reconstruction calculation depends on the accuracy of the measurement with which the calculation is compared.

After the implementation of DS86, measurement of neutron activation in material present at the time of the bombings suggested substantial disagreement between measured and calculated values of thermal-neutron fluence. Additional measurements tended to confirm the discrepancy. Given that the activation measurements showed a uniform trend of divergence from DS86, the first efforts to address the discrepancy were directed at the radiation calculations.

The concerns that the neutron discrepancy raised for the accuracy of survivor doses and for radiation-transport methods led the radiation-physics and biomedical-research programs at the US Defense Nuclear Agency (DNA) to begin in 1988 to address these problems. The group at DNA had a long-standing involvement with work in this field dating back to the end of World War II. DNA had sponsored the development of the transport method that made DS86 possible and had been involved in initial work pointing to the possibility that advances in radiation-transport technology could yield doses for the atomic-bomb survivors different from those calculated with T65D. Three other considerations drew DNA into the problem at this time. First was an appreciation of the importance of the Hiroshima and Nagasaki dosimetry to estimates of risk for radiation-induced cancer. The radiation-protection community, led by the National Council on Radiation Protection and Measurements (NCRP), established that cancer risk was almost the sole effect of low-dose radiation, and was forming the standards of radiation protection. Second were long-standing professional ties to the group at Science Applications International Corporation (SAIC) that implemented DS86. Third was primary sponsorship of developments in radiation-transport methodology at several of the national laboratories at that time. The broad program that DNA then undertook was carried out at Oak Ridge National Laboratory (ORNL), Los Alamos National Laboratory (LANL), Lawrence Livermore National Laboratory (LLNL), SAIC, the Aberdeen Pulse Reactor Facility (APRF), the Department of Energy Environmental Measurements Laboratory (EML), and the National Institute of Standards and Technology (NIST). Between 1988 and 1993, individual projects were undertaken to address many of the aspects of the discrepancy between calculated and measured thermal neutrons. In retrospect, when the project began, no one at DNA or on the Committee on Dosimetry for the RERF expected the problem to be so intractable or to elude resolution to this day.

The first phase of the work concentrated on the validation of the underlying radiation-transport methodology used in DS86. This approach arose out of concern that DS86 was unable to accurately calculate the measured europium or cobalt activation in either Hiroshima or Nagasaki. That inability and inconsistent results in calculating results of controlled-activation experiments at the APRF led to the concern that there might be a fundamental unappreciated flaw in the basic discrete-ordinates (S_n) radiation-transport methodology. This phase of the work—which consisted of numerous projects designed to devise, investigate, and test the impact of possible solutions—was carried out between 1988 and 1991. It included the review of existing neutron cross-section data and evaluations, generation of new neutron cross-section information and evaluations, conduct of proof-of-principle and benchmark experiments, creation of new in situ measurement techniques, and the making of new in situ measurements. The results of this work were to remove the factor-of-10 disagreement between the neutron-activation calculations and measurements in Nagasaki. The results were achieved with the use of new nitrogen and oxygen cross sections for neutrons over a broad energy range, an increase in the number of neutron groups used in the calculation, and the use of different S_n techniques. However, it was possible to confirm agreement in Nagasaki only after new and better measurements were made at greater distances there and a benchmark activation-calculation experiment was completed at APRF (Straume and others

1994). Those changes not only removed the factor-of-10 disagreement between Nagasaki calculations and measurements, but also produced better overall agreement with the atomic-test results and supported the validity of discrete-ordinates radiation-transport methodology. However, the many improvements in measurement and calculation technology that resulted in agreement between the Nagasaki measurements and calculations did not resolve the discrepancy in Hiroshima, which gave rise to the second phase of the DNA effort, directed specifically toward efforts to resolve the Hiroshima neutron discrepancy.

The Hiroshima neutron work took three forms. First, there was an examination of various hypotheses about unexpected ways in which the Hiroshima bomb might have disassembled during explosion. Second, a composite source-term sensitivity analysis was conducted to determine whether any plausible combination of bomb source term and fission-spectrum neutrons could be found that would match all the in situ measured data sets for detonations at a number of bomb burst heights. Third, a calculation exercise was undertaken to "find" a source term for the bomb that would agree with the in situ measurement data. Many of the suggested changes enhanced agreement between the measured and calculated doses of gamma rays and neutrons, but the funding of this program ended before the Hiroshima problem was solved.

THE NAGASAKI DISCREPANCY

The ratio of values calculated by DS86 to those measured in situ for europium and cobalt can be clearly seen to disagree in Figure 4-1 for both Nagasaki and Hiroshima. This comparison, made in 1988, is complicated by the fact that there is only one measurement beyond 1000 m in Nagasaki. Given the undue influence of the 1000 m point on comparisons in Nagasaki, a top priority at that time was to obtain additional activation measurements in Nagasaki. The desire to have isotopes other than europium and cobalt for comparison mandated a search for an isotope present in sufficient abundance to assay and with a sufficiently long half-life to still be plentiful.

Accelerator mass spectrometry (AMS) was proposed as a method to measure small amounts of ^{36}Cl (Haberstock and others 1986; Straume 1988). First, the development and proof-of-principle for this technique had to be undertaken, because this type of chlorine activation assay had not been done before. After the successful development of the AMS chlorine assay at LLNL, measurements were made from concrete cores taken 822, 1187, and 1261 m from the hypocenter in Nagasaki. The measurements were accepted only after they had been shown to be in good agreement with the results of benchmark experiments conducted first at LLNL and then at distance in an open-air reactor at APRF (Straume and others 1994). They constituted a major step in resolving the neutron discrepancy at Nagasaki.

To test the APRF chlorine results, benchmark measurements were also conducted at APRF in 1992 and 1993. These experiments employed a multisphere

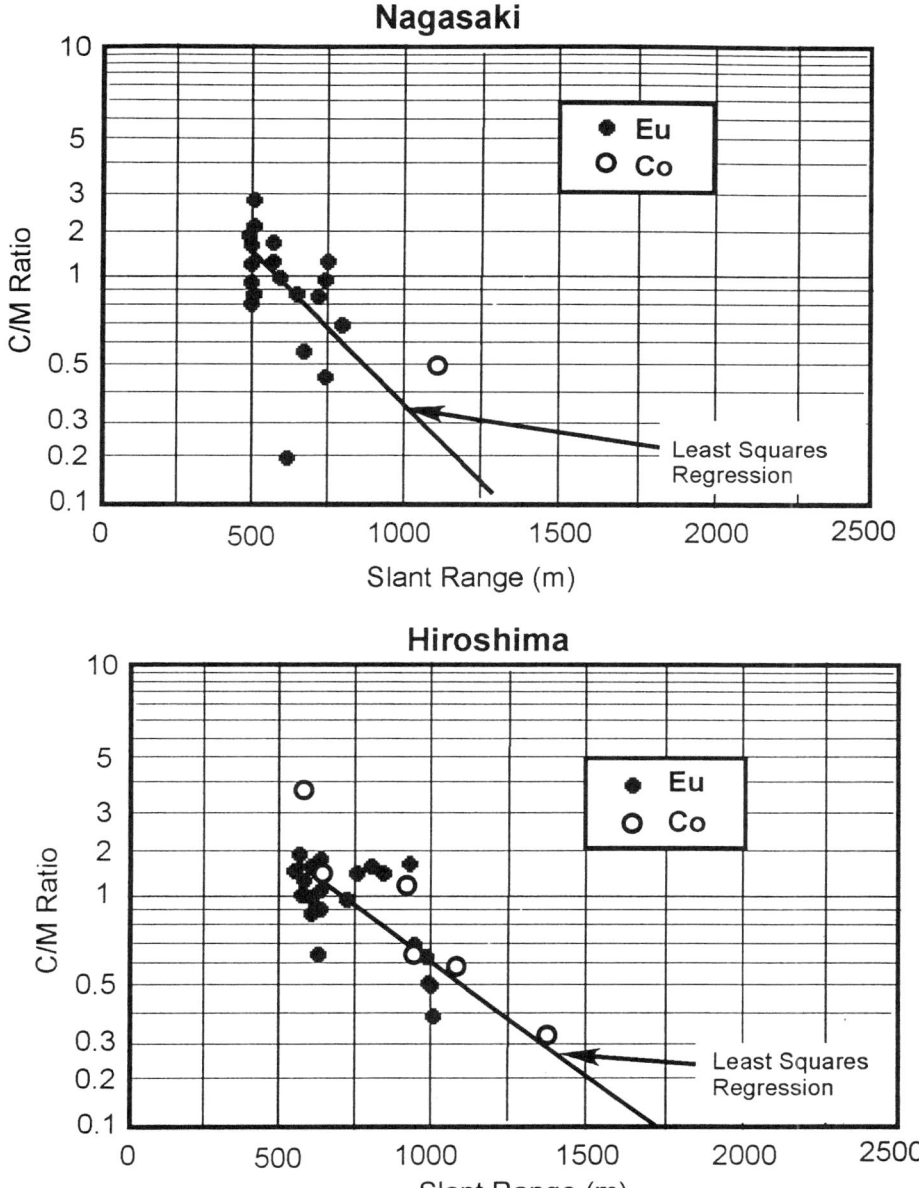

FIGURE 4-1 Nagasaki and Hiroshima C/M for Eu as a function of slant range. 1988.

neutron spectrometer with 12 detectors operating simultaneously at 300 to 1986 m. Neutron fluences and energy spectra were measured by EML using bare, cadmium-clad, and high-density polyethylene BF_3 spheres (Goldhagen and others 1996). The thermal-neutron detectors were calibrated by using the NIST thermal-neutron beam (±12% at 95% confidence). The measurements illustrated the crucial role the water content can play in neutron spectral shape and thermal-neutron production. An increase in ground moisture from 8% to 29% produced a measured $1.46 \times$ increase in thermal neutrons and increased the percentage of neutrons greater than 0.1 MeV from about 27.5% to 34% of the spectrum. Calculation of those and other benchmark fluences was good overall and diverged from the measurements by only 50% at the worst point of comparison.

Thus, the benchmarks support the overall integrity of the radiation-transport calculation methodology used in DS86. The demonstrated ability of S_n to determine benchmark measurements, to calculate nuclear-test results, and to calculate the thermal measurements at Nagasaki makes it highly unlikely that the transport methodology could be responsible for the remaining discrepancy in Hiroshima. That does not rule out the possibility of environments in which thermal neutrons are produced but are not well characterized or not well modeled. Inadequacies in assessing contributions from other elemental isotopes, such as potassium and boron, or the failure to account properly for the water content of the sample could easily create a factor-of-2 error in the thermal-neutron activation in a sample. If that happens, the apparent discrepancies between physical measurement and calculation could create the perception of a greater discrepancy than actually exists in the small fraction of the dose attributed to neutrons as calculated by DS86 for survivors.

Calculation Codes and Cross Sections

Concurrently with the development of new measurements in Nagasaki, changes were made to the discrete-ordinates calculation method (Kaul and others 1994). Some of the changes resulted from developments in the cross-section libraries available for neutrons—for example, refinements in the oxygen and nitrogen cross-sections between ENDF/B-5 and ENDF/B-6. Concomitant increases in computing capacity made it possible to increase the number of neutron groups from 46 to 174 and the number of gamma-ray groups from 23 to 38 (see Table 4-1). This change also increased the number of thermal-neutron groups from one to four, permitting much better resolution of the critical thermal-neutron calculations.

Those changes in combination with a recalculated LANL source term (Whalen 1994; Woolson 1993), changes in the size and details of the air-over-ground geometry of the calculation, the use of newer discrete ordinates calculation and transport codes, as well as changes in the time-dependent source and geometry (see Table 4-1), reduced the calculated thermal activation near the Nagasaki hypocenter by nearly a factor of two and brought the discrete-ordinates calculation into closer agreement with Monte Carlo calculations (Whalen 1994).

TABLE 4-1 Feature-by-Feature Comparison of DS86 and Parameters Suggested in 1993

Prompt-Radiation Methodology

DS86	**1993**
Sources	**Sources**
- LANL 1983 sources	- LANL 1990 sources
- 27 neutron energy groups	- 46 neutron-energy groups, 20 angles
- 20 angle bins	- reformatted to 174 energy groups and continuous angle distribution
Geometry	**Geometry**
- Seven-zone air density profile	-DS86 material compositions and profiles
- maximal radius 2800 m	- continuous vertical density variation
- maximal height 1500 m	- maximum radius 3000 m
- maximal radial mesh 25 m	- maximal height 2000 m
	- maximal radial mesh 25 m
Cross sections	**Cross sections**
- ENDF/B-5	- ENDF/B-6.2
- 46 neutron and 23 gamma-ray energy groups	- 174 neutron – 38 gamma-ray energy groups
- scattering order P3	- Scattering order P3
- custom weighted by region	
Code	**Code**
- DOT-4, 2-D discrete ordinates	- DORT, 2-D discrete ordinates
- first collision source	- first collision source
- 240-direction angular quadrature	- 240-direction angular quadrature
- negative source fixup	- no negative source fixup
- convergence criterion 1×10^{-2}	- convergence criterion 1×10^{-3}
- weighted-difference flux calculation	- theta-weighted flux calculation

Delayed-Radiation Methodology

DS86	**1993**
Time-dependent source	**Time-dependent source**
- neutrons: augmented Maxwellian spectra, E to 2.5 MeV	- neutrons: ENDF/B-6 spectra, E to 8 MeV
- gamma ray: empirical spectra	- gamma rays: ENDF/B-6 spectra
Time-dependent geometry	**Time-dependent geometry**
- line-of-sight optical depth (g/cm^3) from 2-D air density contours	- 2-D air density contours
- contours from STLAMB hydrodynamic code	- contours from STLAMB hydrodynamic code
Transport codes	**Transport codes**
- ANISN: transport in uniform air, 300 time steps	- DORT: transport in 2-D air-over-ground, 12 time steps
- ANISN: transport in hydrodynamically perturbed air	
- morse: fluence perturbation due to the air-ground interference	

As work in upgrading the S_n methods developed and the new ENDF/B-6 cross-sections became available, it was increasingly clear that the first visible resonance in the nitrogen cross section was not properly described. Given that that resonance occurs at about 433 keV, its impact on thermal-neutron calculation could be important. To address that concern, high-resolution measurement of the total neutron nitrogen cross section from 0.5 to 50 MeV was undertaken at the Oak Ridge Electron Linear Accelerator (Harvey and others 1992). Measurements indicated that the size and spin parity for the 433-keV resonance had been substantially misrepresented and that there was some degree of shape distortion even in the broader resonances at higher nitrogen energies. These data and data on oxygen cross sections from Germany (Cierjacks and others 1980; Drigo and others 1976) were used to reevaluate ^{14}N and ^{16}O at low energies with a multichannel R-matrix analysis of reactions in the ^{15}N and ^{17}O systems (Hale and others 1994). The incorporation of the newly evaluated cross sections for oxygen and nitrogen led to a revision of the evaluated nuclear data files to produce ENDF/B-6.2 and changed the calculated scattering in air, creating an inelastic scattering above about 3 MeV and more pronounced forward scattering of neutrons below about 1.5 MeV. Together, these factors resulted in a reduction in the calculated sulfur-activation relaxation length from 220 to 207 m and in closer agreement with tests of Nagasaki-like bombs.

THE HIROSHIMA BOMB AND POSSIBLE OTHER SOURCE TERMS

The combination of the changes in calculation methods, the new chlorine measurement, and the definition and implementation of new oxygen and nitrogen cross sections were all necessary to achieve agreement between the calculations and in situ measurements in Nagasaki. None of those improvements has been implemented in DS86, which has remained essentially unchanged since its inception in 1986. When applied to the calculations in Hiroshima, these same changes have made the agreement between calculations and measurements incrementally better, but have not removed the disagreement as they did in Nagasaki (see Figure 4-2). In Hiroshima, the use of the new nitrogen and oxygen cross sections and the refinement of the S_n techniques have produced calculations that agree more closely with sulfur activation near the hypocenter. The type of chlorine measurements that helped to remove the discrepancy in Nagasaki supports the substantial nature of the discrepancy in Hiroshima (Straume and others 1994).

Given apparent resolution of the Nagasaki discrepancy and the intractability of the Hiroshima discrepancy, DNA began systematically investigating various disassembly and detonation hypotheses that had been advanced to account for the measurements and made a first attempt to derive a source from the measurement data themselves. In Hiroshima, the basic problem is that DS86 does not calculate the level of thermal-neutron activation at distance as is present in materials activated by bomb neutrons. It has been known for some time that postulating a source

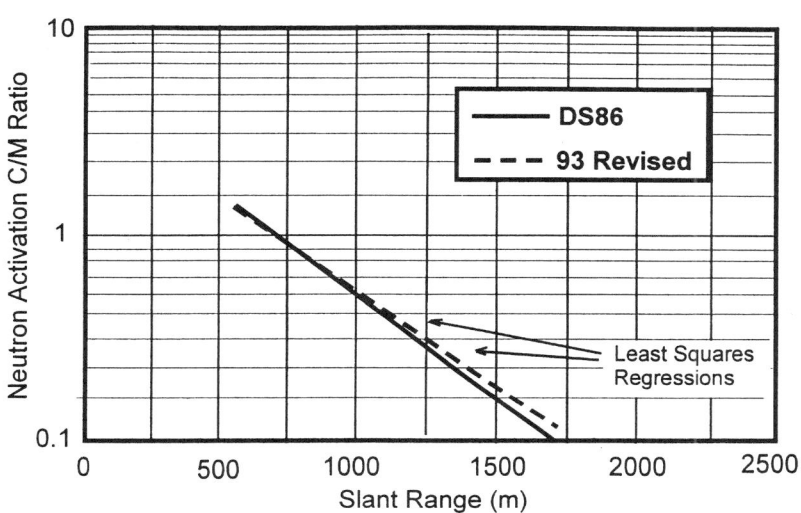

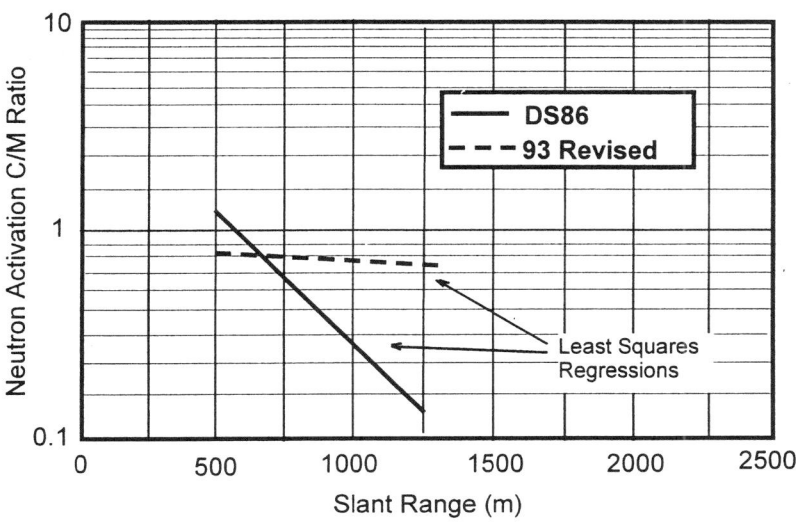

FIGURE 4-2 Nagasaki thermal-neutron activation (1993) revised calculation to measurement ratio as a function of slant range from the hypocenter placed beside Hiroshima thermal-neutron activation (1993) revised calculation to measurement ratio as a function of slant range from the epicenter.

for the bomb that contains more neutrons above 1 MeV in energy could bring about better agreement of the thermal-neutron calculation with measured values. Such a source would produce the greater thermal-neutron relaxation lengths necessary to calculate the activation levels measured at distances of 1 km or more. Because the standard nuclear-weapons codes and conventions do not produce the hard spectrum suggested by the measurements, many hypotheses have been advanced as to the mechanisms by which sufficient high-energy neutrons could have been produced to bring about the measured thermal-neutron activation.

Most of the suggested mechanisms of high-energy neutron release involve abnormal disassembly of the Hiroshima bomb. Such speculation has been fostered by several circumstances peculiar to this bomb. First, as mentioned above, the bomb dropped on Hiroshima is the only one of its type ever detonated. A number of Nagasaki-type bombs have been tested, but there is no bomb comparable with the Hiroshima bomb from which measurements can be derived. Second, the unique design and structure of the Hiroshima bomb has led to hypotheses about structural failure of the bomb case during detonation that would have permitted fast neutrons to escape unmoderated by the thick iron case that surrounded the core. The fact that the bomb case was a steel gun barrel has led to the suggestion that either the bomb case cracked or the tail of the bomb blew off, creating a streaming path for high-energy neutrons to escape.

ALTERNATIVE DISASSEMBLY HYPOTHESIS

Cracks had been reported in some gun-tube assemblies during non-nuclear test firing prior to the building of the Hiroshima bomb (Rhodes 1995). This observation led both Auxier (1991, 1999) and Hoshi and others (1999) to suggest that the Hiroshima bomb could have split from the shock of the high explosive before the bomb reached peak nuclear power providing a portal for the release of fission-spectrum neutrons. They suggested that adding such high-energy neutrons might result in a bomb output spectrum that would better match the in situ neutron-activation measurements. Despite the fact that the case for the Hiroshima bomb was test-fired and did not crack prior to being loaded with the nuclear material (Rhodes 1995), the committee considered the possibility of such an unexpected disassembly and concluded that it was both extremely unlikely and incapable of matching both the fast and thermal neutron data had it occurred.

Given that those alternative disassembly hypotheses could never be tested and that no direct data support or refute nonstandard modes of disassembly of the bomb, the DNA program sought to explore them to test their feasibility. The first step in the process was to review the bomb hydrodynamics and time course with the design group at LANL. The review produced several observations, all of which are incompatible with abnormal disassembly of the bomb. First, major compromise of the case seems improbable because it would have substantially reduced the observed yield of the bomb (Whalen 1994). Second, the course of development of cracks in the case is measured in milliseconds, whereas the neutrons would have escaped from the bomb within microseconds. So the prompt neutrons would have been released before the explosion began to distort the bomb case. Third, small cracks in the case would have to be oriented in exactly the straight-line paths that neutrons take between collisions for the neutrons to escape unmoderated, as Whalen (1994) has pointed out, this is extremely improbable. Fourth, for neutrons arising late in criticality, small to moderate cracks are annealed by the heating of the bomb. These arguments represent the best judgment of the persons responsible for bomb design and evaluation, but in the end it is just the best judgment of experts. To test directly whether adding fission-spectrum neutrons, whatever their origin to the Hiroshima source term, would improve the agreement with measurements, a composite source study was conducted.

The composite-source study was conducted during the summer of 1993 by SAIC (Kaul and others 1994). To test the hypothesis that calculated activation would match the activations measured in Hiroshima better if the source term for the bomb contained more high-energy neutrons, (such as would be the case if the bomb case had cracked prior to nuclear detonation) fission-spectrum neutrons were systematically added to the LANL-calculated Hiroshima output source (similar to the spectrum for inelastic scattering in iron) and compared to the in situ measurements of activation. This study was constrained to determine whether any such combination source could be identified that would agree with the high-energy (sulfur), gamma-ray (TLD), and thermal-neutron activation data. Height of burst for the bomb was also varied to evaluate the influence of height on agreement. In all cases, the addition of fission-spectrum neutrons to the Hiroshima source made the agreement between calculated and measured sulfur deteriorate more rapidly than the agreement between calculated and measured thermal-neutron activation improved (Whalen 1994). This process is strongly governed by the results from the sulfur and thermal-neutron activation and is not strongly influenced by the TLD data, which are relatively insensitive to the assumed spectrum. This can be seen in every analysis that has been undertaken since the inception of DS86, in which there has always been good agreement with TLD data, even in the face of large discrepancies with other measurements. The result from this analysis is that no combination of composite source and height-of-burst could be identified that could simultaneously reconcile the best available 1993 calculation with the TLD, sulfur activation, and thermal-neutron activation data (Kaul and others 1994). The overall implication of all the work done on alternative disassembly hypotheses is that they do not reconcile the Hiroshima calculations and measurements.

The implication is that either the constraining measurements or the leakage for the Hiroshima bomb must be altered. The constraining measurements were the sulfur measurement of fast-neutron activation and the entire set of thermal-neutron activation measurements. The implication of the study for the leakage spectrum of the Hiroshima bomb is that the sulfur and thermal-neutron data can be matched only by tailoring an output spectrum in which there are more neutrons of 2-3 MeV and fewer of below 2 MeV and above 3 MeV, because adding neutrons below 2 MeV would cause the overestimation of the measurements at short distances from the hypocenter and additional neutrons above 3 MeV would disrupt agreement with the sulfur measurements. The next phase of this investigation was to see whether such a source could be identified.

The effort to define a source term that would cause the calculations to match all the in situ measurements in Hiroshima was undertaken by the Mathematical Physics Division of ORNL in 1993 (Rhodes and others 1994). The first calculation essentially repeated composite-source calculations with the actual leakage of the APRF reactor which was placed in a 30-degree horizontal band around the midplane of the bomb. Using the APRF at about one-third of the total neutron leakage calculated for the weapons produced an excellent fit—to within a few percent of thermal-neutron activation data. That solution failed for two reasons: it produced a discrepancy of a factor of 10 with the sulfur data under the bomb, and it required the elimination of most of the conventional leakage (which is a softer neutron spectrum) calculated for the bomb to match the short-range measurements. After several other such failed attempts, it was decided to concentrate on leakage in the energy range above the "oxygen window", which is a large valley in the oxygen cross section at about 2.3 MeV. That was done in two ways. First, 8% of the total neutron leakage of the Hiroshima bomb was concentrated at 2.1-2.7 MeV; this "boost" to the neutron flux was uniformly distributed in a 30-deg band around the horizontal midplane of the weapon. The remainder of the leakage spectrum was adjusted to maintain the number of neutrons, and neutrons above 2.7 MeV were eliminated to match the sulfur data. This output leakage spectrum fit all the data to within 20%. In some of the reports of this work, the leakage spectrum has been referred to as the "pancake source."

One other such source was constructed. It has been referred to as the "funnel-cake source." In this configuration, 31% of the bomb neutrons were concentrated in an upward-directed 30-deg cone, and some of the neutron output above 2.7 MeV was retained. This configuration produced a good fit with all the in situ measurements the largest error was 22%. These source spectra were considered informative but could not be adopted because no explanation for such a neutron release could be posited. The exercise causes one to look carefully at the accuracy of the in situ measurements because large errors in the sparse fast-neutron activation data could permit the boost in leakage to be distributed over a wider energy range above 2.7 MeV. It is interesting to note that either of the "boosted oxygen-window spectra" would have only a very small effect on the gamma doses while placing the neutron dose much closer to the estimates of T65D. On the basis of a preliminary spectral unfolding calculation, Pace (1993) suggested that the output spectrum to match the data would require a strong peak above the oxygen window with source reduction both above and below the window.

In the adjoint calculations associated with all those calculational exercises, Whalen (1994) sees hope that the Hiroshima neutron-activation measurements can be matched with a source spectrum that is softer, not harder. That hope is based on data that show that calculations and measurements of neutron transmission through thick iron (as in the case of the Hiroshima bomb) should be softer than is currently calculated. To follow that lead, the final experiment funded by the DNA program on Hiroshima neutrons was the measurement of neutron transmission through a cross section of the bomb case and tamper. Time-of-flight measurements were made with the LAMPF 800-MeV proton linear acceleration at LANL as a "white" neutron source to measure the transmission of neutrons 0.6-600 MeV.

STATUS OF EFFORTS TO IMPROVE DS86

Following the recommendations of this committee (see Chapter 8), a number of projects have been funded and are in progress in an attempt to refine the calculations of DS86. These include a total recalculation of the output of the Hiroshima bomb, the total reevaluation of shielding models for Hiroshima and the factory workers in Nagasaki, and an evaluation of S_n and Monte Carlo calculations as a method for an adjoint determination of a source for the Hiroshima bomb that will agree with the measurements.

The recalculation of the Hiroshima bomb will be the most comprehensive ever accomplished. It will include a late-time output spectrum that incorporates new iron cross sections and transmission through the bomb case, a new Monte Carlo source term as a function of energy and angle, and a Monte Carlo transport of the new source to the ground accounting for the tilt and heading of the bomb, and the transport of delayed neutrons over time through air created in a new spherical air blast calculation. The new calculations, which will use the latest ENDF/B-6 cross sections, will be compared with source and DORT calculations. If it is necessary for a satisfactory level of agreement, an adjoint-to-source and forward to free-in-air kerma Monte Carlo calculation will be performed.

The next kind of work recommended by this committee is an examination of the adequacy of shielding models in DS86. Two efforts to follow this recommendation are under way. The first is an examination of the nine-parameter, globe, and terrain shielding in Hiroshima. This effort will identify required changes in how shielding is handled in DS86. The second effort is the improved modeling of the shielding environment for the factory workers in Nagasaki. The biological dosimetry for these workers indicates that the shielding could be in error. With better accounting for the shielding provided by these structures and the heavy machinery they contained, the RERF dosimetry system can determine the dose for these workers better.

The work to improve the radiation-transport calculations should be completed as quickly as possible. Any improvements derived from the work and other improvements in radiation cross sections and transport methodology achieved since DS86 should be fully implemented in the RERF dosimetry system when they have been reviewed by US and Japanese senior review panels, fully documented, and approved.

5

Biological Dosimetry at RERF

Unlike DS86, which provides a calculated estimate of a person's organ dose, biological dosimetry estimates dose by evaluating a marker left in the tissue by radiation exposure. The marker can be any of a number of signatures or can be evidence of radiation damage itself. These markers are, in general, a function of the magnitude of an exposure, and they have for some time offered the promise of specific dose determination for each person exposed to ionizing radiation.

Efforts to determine biological doses of survivors in Hiroshima and Nagasaki began at RERF in 1968, shortly after cytogenetic dosimetry was first shown to be useful in radiation-dose reconstruction. However, not all markers of radiation exposure persist; the body repairs or deletes damage that is used as an indicator of exposure. For instance, the most reliable and best documented bioindicator of radiation exposure—the number of dicentrics in peripheral lymphocytes—persists for only about 3 years or less; although dicentrics are excellent for recent accidents, they are not useful for dose reconstruction in the Japanese survivors more than 50 years later. Therefore, biological dosimetry at RERF has had to turn to stable aberrations or other markers that persist for a lifetime. Until very recently, the assay of such persistent markers has been cumbersome and time-consuming, and this has resulted in biodoses for only a small fraction of the people in the Life Span Study (LSS) cohort.

Efforts to find faster assays and more reliable biomarkers of dose have led to RERF attempts to do work by conventional staining analysis and G-banding for cytogenetic aberrations, glycophorin-A assays, electron-spin resonance (ESR) of calcium-tissue samples, and fluorescence in situ hybridization (FISH) for the detection of reciprocal translocations in human chromosomes (See Table 5-1).

Table 5-1 summarizes a number of markers that have been used in biological dosimetry and compares their attributes. Most of these methods have been explored

TABLE 5-1 Comparison of Biomarker Attributes (Straume and Lucas 1995)

Biomarker	Human *in vivo*	*in vitro*	Animal Model	Inter-Person Variation	Persistence Post-Exposure
Translocations[a]	Yes	Yes	Yes	Low	0–lifetime
Dicentrics	Yes	Yes	Yes	Low	0–6 mos.
Micronuclei	Yes	Yes	Yes	High	0–6 mos.
HPRT[b]	Yes	Yes	Yes	Medium	1 mo.–1 yr.
GPA[c]	50%	No	No	High	6 mo.–lifetime
TCR[d]	Yes	No	No	High	1 mo.–2yrs.
HLA[e]	50%	Yes	No	?	1 mo.–1 yr.
SCEs[f]	Yes	Yes	Yes	?	0–6 mos.
DNA Adducts	Yes	Yes	Yes	?	0–6 mos.
Protein Adducts	Yes	Yes	Yes	?	0–6 mos.

[a] Reciprocal chromosome translocations.
[b] Hypoxanthine phosphoribosyltransferase assay.
[c] Glycophorin-A somatic mutation assay.
[d] T-cell antigen receptor mutation assay.
[e] Human leukocyte antigen mutation assay.
[f] Sister chromatid exchanges.

at RERF and for various reasons, some obvious, many have not been suitable for the type of dose determination required.

Despite the promise of FISH and other new assays to produce the assay speed that would allow biodose determination for large numbers of survivors in the RERF study cohort, fewer than 4000 of the over 86,000 in the LSS population currently have biologically-determined doses. Most of these individual doses are derived from the very labor-intensive work with G-banding or from ESR. In addition to the biomarkers compared in Table 5-1, individual exposures to x or gamma radiation can be determined by using electron spin resonance to detect the physical signature left in tooth enamel by irradiation. The ESR signal is proportional to dose between 0.1 and 10 Gy and is independent of photon energy above 200 KeV. As such, ESR signals obtained from the teeth of survivors are good indicators of the doses from the high gamma ray energies released by the A-bombs. This technique has been used successfully to reconstruct dose in some radiation accidents (Pass and others 1997) and at RERF for a limited number of survivors (about 60). The comparison in Figure 5-1 of the dose derived from ESR and chromosome translocations for the same Japanese survivors shows excellent agreement within the uncertainties of these types of assay.

Given the small numbers of biodoses available and the labor currently involved in producing such individual doses, biological dosimetry is not currently a viable alternative to an overall dosimetry system that produces a dose estimate for all of the members of the LSS. Although only a small number of biologically-derived doses for people in the LSS cohort have been determined, these estimates

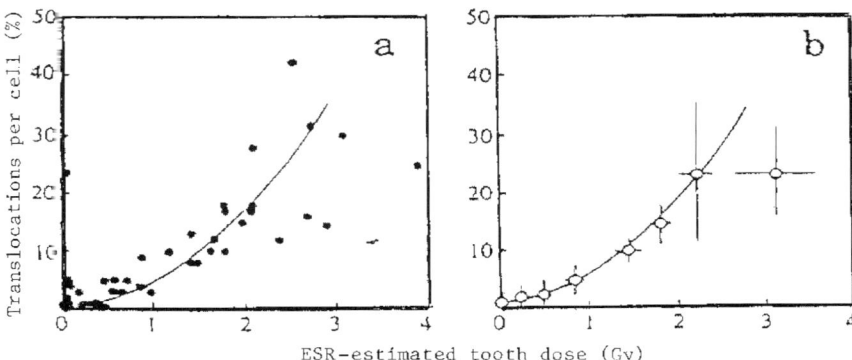

FIGURE 5-1 Translocation frequency of lymphocytes from 41 tooth donors measured by conventional Giemsa staining plotted against ESR-estimated gamma-ray dose for lingual portions of molars. (a) Individual data; (b) grouped data (each point consists of five individuals) (Nakamura and others 1998).

are important in finding out whether there are problems with DS86. As an independent individual dose assessment, biological dosimetry can help to characterize problems and suggest work that must be done to make any successor dosimetry system an accepted source of dose information. For example, areas where biological indicators of radiation dose and the doses derived from DS86 disagree can be looked for. Comparisons of specifically selected subsets of the LSS cohort can yield important evidence to support or refute discrepancies between DS86 and in situ physical activation measurements. Analysis of the trends in such comparisons can be vital clues as to the nature and direction of discrepancies. Two issues of DS86 accuracy in predicting dose are addressed with biological data: the possible underestimation of neutron doses in Hiroshima and the possible overestimation of gamma doses to workers in the Nagasaki torpedo factories.

Differences Between Hiroshima and Nagasaki

The first source of biological data, which has been interpreted by some as evidence of underestimation of neutron dose in Hiroshima, comes first from epidemiology, and second from biological dosimetry. The most recent analysis of cancer risk data from the LSS cohort suggests that excess cancer rates are higher in Hiroshima than Nagasaki (Pierce and others 1996). Interpretation of that observation is a contentious issue, and it can be attributed to various factors. Some have pointed to it as evidence that the more biologically injurious neutrons are the reason for a higher excess rate in Hiroshima than in Nagasaki, where the neutron doses are generally conceded to have been lower (see Chapter 7 for a discussion of this issue). The epidemiological observation is consistent with the biological dosimetry for the two cities. Stram and others (1993) reported a greater number of stable chromosomal aberrations in the lymphocytes of members of the LSS cohort in Hiroshima than for the same DS86-estimated dose in Nagasaki. Analyses and explanations of these findings have looked to differences in the populations of the cities and to overall underestimation of the Hiroshima dose as possible explanations, but they have not been able to rule out the possibility that the difference is attributable to an underestimation of the neutrons in Hiroshima DS86 doses.

Recent analysis of stable chromosome aberration data obtained from Geimsa-stained cultures from approximately 3000 survivors in Hiroshima and Nagasaki confirms there is a statistically significant difference in the number of chromosome aberrations seen in the two cities at any given dose (Nakamura and Preston 2000). Survivors in Hiroshima had an average of 6.6% aberrant cells per sievert of exposure; whereas Nagasaki survivors had 3.7% aberrant cells per sievert (see Figure 5-2).

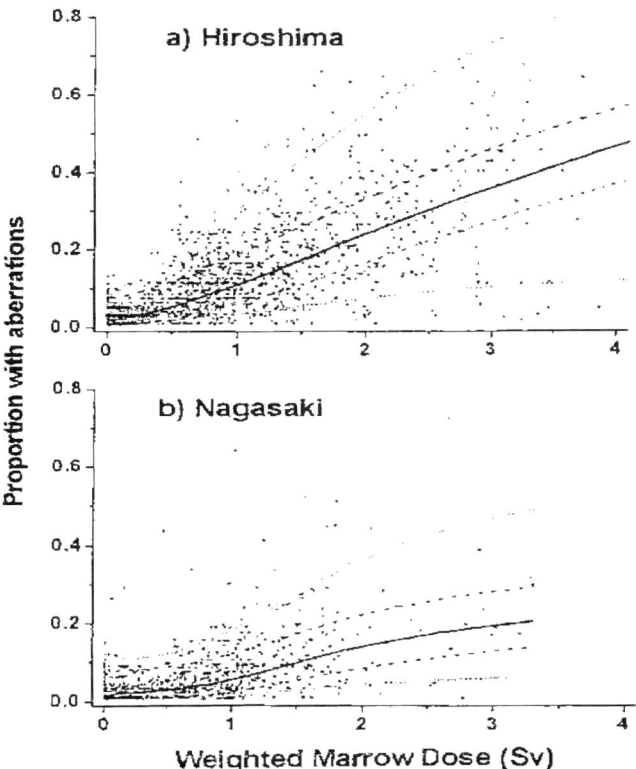

FIGURE 5-2 Scoring efficiency-adjusted proportion of cells with aberrations versus dose. The solid curve is a nonparametric estimate of the dose-response function (Kodama and others 2001).

The Nagasaki Factory Workers

A comparison of aberrations reveals inconsistencies between the number of aberrations and the shielding calculated for survivors in DS86. This is especially true for people located in the large metal torpedo factories in Nagasaki. These survivors had lower numbers of aberrations than individuals exposed to the same dose calculated by DS86 in Nagasaki houses, suggesting that DS86 over-estimates the doses to these people by approximately 40%. This observation agrees with previous ESR work based on the examination of tooth enamel.

When gamma-ray doses for the workers in the Nagasaki torpedo factories are determined with ESR (Nakamura 1999, 2000), they are 40–60% less than those estimated by DS86 (Preston 1999). The fundamental difference between this cluster of large sheet-metal factories filled with heavy machinery and the typical DS86 radiation-shielding case—a wood frame residence—has led to concern that DS86 does not adequately account for the shielding of the factory workers. If that is the case, it causes the current dosimetry system to overestimate the gamma-ray dose received by the roughly 800 workers in the factories. This does not seem like a large percentage of the 86,000 people in the LSS cohort, but such an overestimation becomes important when one recognizes that the 800 are about 40% of the survivors who received 0.5–2.0 Sv in Nagasaki. They are a large percentage of the moderate to high-dose survivors in Nagasaki and are therefore crucial to any accurate risk analysis that involves the Nagasaki data. Thus, it seems vital that any future work to upgrade the dosimetry system for RERF contain a state-of-the-art method of handling the shielding in complex structures. This matter is presently under investigation. Obviously, without biological dosimetry studies on these workers as a source of independent dose determination, the magnitude and direction of the possible miscalculation of the doses to the factory workers would not have been seen.

The limited amount of biological dosimetry that is available for the survivors in the RERF cohort has proven to be extremely useful. Biodoses have served to confirm a potentially significant difference between the two cities, which if substantiated could provide valuable information about the effect of the different qualities of radiation in the two cities (see Chapter 7). ESR and stable chromosome aberrations have highlighted important potential inadequacies in the shielding models currently used in DS86. Given the vital role that biological dosimetry played in this specific case, it is prudent to expand the number of cases in which it is used. Techniques like FISH and ESR that are in place at RERF could be extremely useful in providing an independent dose assessment for complex and uncertain shielding configurations. With careful controls for background aberration rates, FISH could potentially resolve the persistent differences in aberrations and effects observed for the two cities. ESR is equally valuable in cases where the gamma dose needs independent corroboration. Biodosimetry cannot currently replace the dose calculated for the entire LSS; but it does provide an invaluable source of dose evaluation for testing the integrity of DS86 or any subsequent dosimetry system.

6

Uncertainty in DS86

It had been part of the planning for the DS86 to produce a complete uncertainty assessment. However, in spite of the aspirations and plans of the working group and the Committee on Dosimetry for the RERF, a complete assessment of uncertainties did not materialize; and at the time of publication of DS86, a complete uncertainty assessment had not been carried out. A temporary assessment of uncertainty in the DS86 kerma estimates was performed. Chapter 9 of the DS86 report (Roesch 1987) describes the DS86 method of computing the uncertainty in the dose of an individual survivor. Uncertainties associated with various components of the system were also described.

The temporary uncertainty assessment yielded estimated fractional standard deviations (FSDs) for various key parameters and preliminary crude estimates of the correlation among system components. It then combined uncertainties by using standard nonparametric methods that were valid as long as the component FSDs were relatively small (<40%). It made no assumptions regarding probability distributions for these parameters, and it relied heavily on estimates of uncertainty and correlation coefficients given by the various authors of the model components. Additional uncertainty estimates were based mainly on the judgments of the DS86 authors. The correlation estimates, in particular, were based on very little analytic support. Some of the preliminary estimates of FSDs were quite tentative, particularly for the neutron component, in which case the apparent disagreement between measured and calculated thermal-neutron activation suggested a possible unresolved problem in the neutron-transport models.

The review of DS86 report by the National Research Council in 1987 (NRC 1987) recommended that a rigorous uncertainty analysis be undertaken with improved uncertainty input values for each aspect of the dosimetry system. The review stated "that the full usefulness of the DS86 system could not be realized until the uncertainty in the organ dose estimates have been properly codified and incorporated into the DS86 system. Quantitative information on uncertainty as a function of distance is an important parameter in the analysis of radiation effects." The present committee was presented with a first draft of such an analysis (Kaul and Egbert 1989). Although it is more complete and rigorous than the preliminary analysis in the DS86 report itself, it has not undergone formal peer review. Furthermore, some of the assumptions and values assigned to various parameters are debatable, as are some of the estimates of correlation between various parameters.

On the basis of temporary uncertainty assessment in DS86 and the draft report mentioned above, errors in kerma to a specific organ of a single survivor have been estimated to be represented by an FSD of about 25–40% (NCRP 1997). The two analyses suggested that the largest contribution to uncertainty is house-shielding, which depends primarily on the location of the survivor in the house and his or her shielding conditions. The committee feels that the uncertainty estimates are too low. Technical improvements in the transport models alone—cross sections, energy bin structure, and delayed neutron transport (see Chapter 4, Kaul 2000)—have suggested an increase of about 30% in neutron kerma at Hiroshima at a 1300-m slant range. As discussed elsewhere in this report, the apparent discrepancy between predicted and measured neutron fluences suggests that the uncertainty in neutron kerma for any given person is larger than in the present uncertainty assessments. It is also possible that some unknown sources of error have not been considered in the uncertainty assessments and that the extent of others has been underestimated because of a lack of information. As pointed out in NCRP report 126 (NCRP 1997), uncertainty analysis of the atomic-bomb survivor data that accounts fully for all sources of error in dosimetry would be very difficult even if all the sources could be fully characterized. Recent reevaluations of potential errors in shielding assignments suggest that gamma doses are more uncertain than indicated in the current assessments (Kaul and Egbert 1989). Evidence from biodosimetry data and the Nagasaki factory-worker effects history indicates further that the preliminary DS86 estimates of uncertainty in individual gamma doses are lower than they should be, even though the overall agreement between DS86 calculations and TLD measurements for gamma rays is very good.

Uncertainty in DS86 can be divided into three types: systematic uncertainty that would affect estimates of doses to all people or groups at about the same distance in the same manner, random errors resulting from the method, and random errors resulting from the input data. The random errors would affect estimated doses to individuals independently.

SYSTEMATIC UNCERTAINTY

System components that can contribute to systematic error are device yield, radiation output, height of burst, location of hypocenter, air density, air and soil moisture, transport methodology, fission-product radiation, shielding methodology, organ-dose calculation methodology, and transport cross sections.

• The evaluation of the yield of the Hiroshima device and its uncertainty should be improved. The uncertainty in the Hiroshima yield is quite high—±2 kT (CV = 10%), and recent reevaluations indicate that a higher yield than the 15 kT used in DS86 could be more appropriate (Kaul and Egbert 1998). However, this reevaluation relies heavily on the comparison of measurements and calculations of the ^{32}S activation near the epicenter and might be modified on the basis of the results of forthcoming ^{63}Ni measurements. (The improvement in air cross sections and energy bin structure lower the calculated ^{32}S activation compared with that using the original DS86 method). The number of neutrons escaping the bomb casing is also uncertain and should be reevaluated. The estimated confidence in the Nagasaki yield is much higher than that in the Hiroshima yield (CV = 5%) (Roesch 1987).

• The uncertainty in the height of burst at Hiroshima was estimated to be ±15 m (99% CI) (Roesch 1987). However, a larger error than reflected by this uncertainty could account for some of the discrepancy in measured and calculated thermal-neutron activities close to the epicenter, so this should be reevaluated for any new dose system. The uncertainty in the height of burst at Nagasaki is estimated to be only ±10 m (Roesch 1987). Errors in height of burst translate into uncertainties that vary with distance, with the greatest impact close to the epicenter. Because an error in height of burst at Hiroshima could explain some of the apparent bias observed in comparisons of thermal activation near the epicenter, height of burst at Hiroshima should be reexamined.

• The location of the hypocenter is believed to be a relatively minor contributor to overall uncertainty.

• The fission product and thus gamma and neutron delayed sources were calculated on the basis of thermal neutron (reactor) fission yields, and this might have resulted in an error in the radiation-source terms of about 10% for neutrons and 5% (CV) for gamma rays. Shortly before publication of DS86, the energy spectrum of delayed neutrons used in DS86 was found to be harder (that is to contain more-energetic neutrons) than estimated, and the delayed-neutron spectrum given in the DS86 publication is thus not the actual spectrum used in the current official DS86 system (Egbert 1999). Later improvements in the delayed-neutron source and transport indicate that the contribution from delayed neutrons was significantly underestimated in DS86 (Egbert 1999). The revised delayed-neutron transport, which has not yet been implemented in DS86 (see Chapter 4), would tend to reduce the uncertainty associated with the delayed-neutron contribution to kerma, according to the observed improvement in the agreement between activation calculations and measurements at Nagasaki and for NTS tests of devices similar to the Nagasaki device. The delayed neutrons were a relatively small contributor to the neutron kerma at both Hiroshima and Nagasaki (<5%) even after the above improvements, but they contributed about one-third of the neutron activation at 1500 m (ground range) at Nagasaki and 8% at Hiroshima according to the original DS86 calculations (Roesch 1987). The revised delayed neutron contribution to activation should be considered in comparing calculated and measured thermal activation.

• The prompt-neutron output from the Hiroshima device is also estimated to have an uncertainty (CV) of about 10%. However, both the number and energy distribution of the neutrons from the Hiroshima source might be considerably more uncertain, as discussed below. The new calculation of this source being carried out at LANL (see Chapter 4) should provide an improved estimate of uncertainty for the total radiation output and for the energy and angular source spectra.

• Errors in air and soil density and moisture content can affect the transport of low-energy neutrons in particular but would probably have only a small impact on the kerma estimates—CV about 5%, according to Kaul and Egbert (1989). However, these errors might have a substantial impact on the calculation of thermal-neutron activation for some locations. The DS86-calculated thermal and epithermal neutron fluences vary by as much as about 25–50% as the altitude increases from 1 m to 25 m (see Chapter 3). A sensitivity analysis of the effect of the uncertainty in these values on the calculated low-energy component of the fluence at various distances and heights should be included in the uncertainty assessment. The variations in low-energy fluence and their possible impact on the comparison between measured and calculated activation are discussed in more detail in Chapter 3.

• Errors and limitations in the shielding and organ-dose methodology (forward-adjoint fluence coupling) could have had a relatively small impact on the estimated kerma (around 5–10% CV), as discussed by Roesch (1987).

• The uncertainty in kerma and activation due to uncertainty in air cross-section values increases with distance, particularly for the neutron component (Lillie and others 1988). The effect on the neutron-kerma uncertainty was estimated to be about 15% (CV) at 1500 m; the prompt and secondary gamma CVs were estimated to be about 3% and 6%, respectively. The uncertainty in nitrogen and oxygen cross sections and improvements in the transport code energy bin structure have been extensively investigated since 1986 and appear to have a substantial impact on the calculated kerma in air. Kaul and Egbert (1989) estimate a CV of about 15% for both Hiroshima and Nagasaki for the uncertainty in the neutron kerma due to uncertainty in the new cross sections (see Chapter 4).

RANDOM ERRORS RESULTING FROM METHOD

Components that can result in random error are uncertainties in the assumed survivor-shielding and organ-dose model.

• The shielding assignment was estimated to contribute substantially to total uncertainty (Roesch 1987). Estimates of terrain shielding at Nagasaki and the model used to estimate the shielding of factory workers also have large uncertainty. Kaul and Egbert (1998) estimated that errors in the shielding calculations for the Nagasaki factory workers could be responsible for an uncertainty of a factor of 2 in the organ-dose estimates. The global and nine-parameter models used to estimate shielding contribute to overall uncertainty. Uncertainty in shielding was estimated to be the largest contributor to overall uncertainty in the total-kerma estimate—CV about 20–40% (Kaul 1999). The biodosimetry data are consistent with a large random uncertainty of about 45% (Sposto and others 1991; Kaul and Egbert 1998) in DS86 dose assignments (see Chapter 5).

• The uncertainties in the organ-dose model are believed to be relatively minor contributors to overall uncertainty.

RANDOM ERRORS RESULTING FROM INPUT DATA

Random error due to uncertainty in input data arises in connection with survivor location, shielding, and survivor orientation. Survivor location and shielding description were estimated to have the greatest contribution to total random uncertainty, primarily because of uncertainty in survivor recall. Kaul estimated that the overall uncertainty in gamma-dose estimates due to uncertainty in survivor recall was around 15% (CV) (Kaul 1999).

DS86 kerma estimates might be even more uncertain because of additional systematic bias in the methodology (Kaul and Egbert 1989) that would affect doses to some or all subjects nonrandomly. Comparisons of activation measurements with calculations of activation based on DS86-calculated fluences indicate additional bias in kerma estimates due to a systematic bias in free-field neutron transport. This systematic bias, unlike most of the systematic errors discussed above, appears to depend on the distance from the epicenter. The apparent discrepancies between measured and calculated neutron activation close to the epicenter and at great distances in Hiroshima, and perhaps also in Nagasaki, imply that such bias exists at least for the thermal and epithermal components of the neutron radiation field; at great distances, this bias is large enough to imply that neutron kerma is also affected. Fluctuations of only about a factor of 2 can occur in the low-energy fluence that produces the activation, because the thermal and epithermal neutrons that contribute to the activation of surface and near-surface samples have a small range in air. (About 30-40% of the total calculated fluence is in the thermal bin; even if all the neutrons were thermalized in the sample, the activation would be increased by at most a factor of 2-3.) A large excess (a factor of 2 or more) of thermal and epithermal neutrons at any distance can arise only if there is down-scattering of higher-energy neutrons in the immediate vicinity of the sample. The apparent discrepancy of about a factor of 3-5 that remains at Hiroshima (see Chapter 3) suggests a bias in the relative number of higher-energy neutrons emitted from either the device or the fireball or in the number of higher-energy neutrons able to survive long-range transport in air (the effective relaxation length). Comparisons of calculated and measured flu-ence from the Aberdeen fission-reactor experiment suggest that uncertainty in trans-port in air based on DS86 was probably relatively small (less than 50%) and thus cannot account for much of the observed discrepancy. However, many of the com-parisons were made with improved transport codes and cross sections rather than with the actual DS86 models and cross-section values (see Chapter 4).

The comparison of the TLD data and the DS86 calculations suggests the pos-sibility of a smaller discrepancy (20% or less; see Chapter 2) in Hiroshima in the gamma fluence and thus gamma kerma. However, the possible discrepancy in the gamma fluence might be directly or at least partially related to the apparent dis-crepancy in the neutron fluence. The preliminary uncertainty assessment in DS86 recognized the apparent discrepancy between the Hiroshima neutron calculations and measurements, and was one of the reasons for the decision to defer a final un-certainty assessment until its cause was resolved. If the improvements in neutron-transport calculations since DS86 described in Chapter 4 are applied, it appears that the neutron-activation measurements and revised calculations in Nagasaki agree fairly well. However, as discussed in Chapter 4, the activation-measurement re-sults for Nagasaki at low activity (at neutron fluences corresponding to those in Hiroshima at distances of 1000–1500 m) are few, and one cannot exclude the pos-sibility that a similar, although perhaps smaller, discrepancy also exists in Nagasaki.

It is important to recognize that the disagreement between neutron-activation measurements and calculations in Hiroshima might result from a combination of systematic errors (including errors in yield, height of burst, and energy spectrum of source neutrons), errors in delayed-neutron source and transport, measurement errors (including background subtraction errors), and activation-calculation errors.

An additional possible location-dependent discrepancy might also be due to the shielding models used in DS86. The failure to account for shielding by other than immediately adjacent structures could have resulted in underestimating the shielding at great distances. A benchmark study carried out by SAIC for subjects whose doses were calculated with the globe model (Kaul and Egbert 1989) indi-cates that DS86 has a tendency toward too low a dose, particularly for the gamma-ray component. A more rigorous modeling (which is funded) of the survivor shield-ing in both Hiroshima and Nagasaki that includes taking account of adjacent structures and terrain features should reduce the uncertainty in these components of the dose system considerably. Similarly, the shielding estimates for the factory workers in Nagasaki are uncertain and the DS86 doses could be biased too high (by as much as 50%), to judge from chromosomal aberration data and ESR data (see Chapter 5). A new study is under way to model the shielding of these work-ers much more rigorously than was done in DS86. An additional segment of the population for which a potentially large bias in the doses might have resulted is the survivors in Hiroshima who resided behind the hill known as Hijiyama. The two-dimensional air-ground model used in DS86 might reflect the scattering at low ac-tivity inaccurately because of terrain variations. Kaul (1999) estimated that the overall uncertainty in gamma dose estimates due to uncertainty in survivor recall are on the order of 15% (CV), which is less than the estimate Jablon once made based on T65D (Jablon 1971).

It is clear that any modification of DS86 should be accompanied by a comprehensive uncertainty analysis that treats all the possible sources of error discussed above and combines them properly, accounting for known correlation, to provide a reasonable estimate of uncertainty in the neutron and gamma components as a function of distance, location, and details of exposure (shielding). The analysis should carefully distinguish between random and systematic error because random error can result in a statistical bias in risk estimates that are based on dose estimates (NCRP 1997). Such an analysis is much more feasible now than in 1986 and 1989 because additional information is available on the possible sources of uncertainty, as are faster computers that will allow benchmark and sensitivity studies of the various model components. The analysis should include the sensitivity of the calculated kerma to the uncertainty in the various cross sections used in the model, to small changes in the energy and angular distribution of the radiation emitted from the device and the fireball, and to the use of a two-dimensional air-ground model, as opposed to a model that reflects the varied topography of the city (this might be particularly important for Nagasaki, where the terrain is very irregular). The uncertainty analysis should clearly indicate which kinds of uncertainty are possibly underestimated and what sources of potential error have not been considered because of insufficient information. A comprehensive biodosimetry analysis can also provide additional information and identify bias in the dose-system results for some populations.

It would be instructive to estimate probability distributions for the most important contributors to uncertainty. These could be used in a detailed stochastic analysis of the distributions of possible doses for selected representative exposure scenarios, including a wide range of distances from hypocenter and shielding configurations. Such an analysis would be much more informative than the simple estimation of total coefficients of variation based on combining fractional standard deviations, and it would provide more-reasonable estimates of the confidence limits on the dose estimates for various representative exposed subjects.

Finally, a comprehensive uncertainty analysis should include a rigorous estimate of the uncertainty in the calculations of neutron activation and TLD dose used to confirm the transport models; additional uncertainties that are not included in the uncertainty model for kerma come into play in these calculations. These include uncertainties in activation cross sections, attenuation in samples, sample orientation, backscattering effects, and the shape of the calculated neutron spectrum at low energy; the latter uncertainty is due to the limited energy bin structure of the transport models. As seen in Chapter 3 in Table 3-1, DS86 contains only a single-thermal energy bin that includes all neutrons below 0.4 eV. An average thermal cross section must be applied to the uncalculated spectral distribution of neutrons below 0.4 eV, and group-averaged cross sections must be used for the limited number of epithermal-energy groups. Note that the latter does not effect the kerma calculations, which are dominated by high-energy radiation. Although uncertainty in the calculation of activation is unlikely to account for a sizable fraction of the observed neutron discrepancy, it might have an impact on the comparison of thermal-activation measurements versus calculation. The discrepancy discussed above in the shielding model with respect to intervening buildings may also be relevant to the thermal neutron activation calculations at great distances in creating additional scattering and thus a possibly greater fraction of lower-energy neutrons; this is discussed in more detail in Chapter 3.

The apparent discrepancy between measured and calculated activation and TLD measurements might not be completely resolved by a revised dosimetry system that incorporates improved source and transport calculations. Nevertheless, a thorough uncertainty assessment can provide credible estimates of the confidence limits on the major component of the dose, the gamma rays, and for the lesser neutron component for representative exposed subjects. It should increase the confidence of both the scientific community and the Japanese population in the validity of the new dose system.

7

Implications for Risk Assessment

This report is generally concerned with the dosimetry of atomic-bomb survivors based on calculation, measurement, and biological approaches. It is appropriate to discuss some factors that could be relevant to the use of the dosimetry in, for example, risk assessment. Even in DS86, neutrons could play a part; if neutron fluence is higher in Hiroshima, as the discrepancy implies, the part played by neutrons could be larger. This section explores that aspect of the matter in an illustrative, rather than a definitive, way. It is not aimed specifically at risk estimation, which must embrace all aspects of epidemiological and dosimetric factors in both Hiroshima and Nagasaki.

CONCERNS ABOUT DS86

The apparent discrepancies between calculations and measurements of thermal neutrons in Hiroshima have led in recent years to concern that there might be substantial biases in risk estimates derived from the atomic-bomb survivor experience. It needs to be emphasized here that that concern is largely unfounded. Although any unresolved aspect of the atomic-bomb dosimetry adds to the uncertainty in risk estimates, the neutron discrepancy has only minor implications for the assessment of risks to survivors themselves. In line with the principal aim of the studies at RERF—the elucidation of the health effects among the survivors—it is reliably known from 5 decades of epidemiological follow-up what the specific radiation exposures in Hiroshima and Nagasaki have done and what past and continuing risk they have posed for the survivors (Thompson and others 1994; Pierce and others 1996). The increased cancer rates have been thoroughly studied in their dependence on distance from the hypocenter and on shielding. Estimates of risk and probabilities of causation are not at issue. They are largely independent of the eventual resolution of the neutron discrepancy.

The unresolved neutron problem is related to a second task of the investigators at RERF. The question is not *what* the radiation in Hiroshima has done to the survivors, but *how* it has done it. The issue is how much of the observed effect can be attributed to the different radiation components—the major absorbed-dose contribution by the gamma rays and the minor absorbed dose contribution by neutrons. Attempts to settle that question are not required primarily for improving the risk assessment for the survivors. They are needed to ascertain whether and how the observations in Hiroshima can be applied to the estimation of risk in populations exposed to radiation that differs from that in Hiroshima, e.g., consisting of a different mixture of gamma rays and neutrons or in being free of neutrons. As already said, such conclusions would be important to the people of the entire world. Therefore, although this report's evaluation of the discrepancies between measurements and DS86 calculations should not, and does not, depend on how the resolution of the discrepancies might affect estimates of radiogenic risk, it is appropriate to consider in a preliminary way whether modifying the dosimetry system to reduce the discrepancies will greatly affect risk estimates.

To be relevant to risk estimation, discrepancies between calculations and measurements of neutron fluence in DS86 must occur in a dose range where health effects due to the radiation exposure have been ascertained, and the neutron doses in this range must be large enough to contribute substantially to the observed effects. There is critical interest in the neutron fluences at distances from the hypocenter between 1000 m and 1500 m, which correspond in Hiroshima to mean organ doses of 0.2–2.0 Gy (Roesch 1987). The fluence values at 2000 m and those at less than about 1000 m are informative inasmuch as they can help to substantiate the values in the region of interest, and the doses below 0.2 Gy could become increasingly important since direct examination of the risk from doses below this level has been considered by Pierce and others (1996) and Pierce and Preston (2000).

On the basis of the earlier dosimetry system, T65D, it had been surmised (Rossi and Kellerer 1974; Rossi and Mays 1978) that neutrons were responsible for a substantial fraction of the late health effects observed in Hiroshima. Support for that idea declined when DS86 specified considerably lower neutron doses in Hiroshima. It was then concluded that the neutrons are, even in Hiroshima, a minor potential contributor to the observed health effects and their role, although uncertain, is not critical for risk estimation. In later analyses, the neutrons were therefore accounted for crudely by applying a weighting factor of 10 to their absorbed-dose contribution. The sum of the gamma-ray absorbed dose and the weighted neutron dose was termed *weighted dose* and was expressed in sieverts. That approach seemed to confirm the relative unimportance of the neutrons in DS86 for risk evaluation. However, a more quantitative assessment of the actual doses is required to appreciate the situation. As we shall see in the illustration given here, a more precise detailing of the neutron and gamma doses reveals a potentially greater role for neutrons, at a total dose of about 1 Gy, than previously envisioned.

THE NEUTRON/GAMMA-RAY DOSE RATIO

The diagrams in Figure 7-1 represent the neutron/gamma-ray dose ratio, (the ratio of average neutron absorbed dose to gamma-ray absorbed dose) plotted against the total absorbed dose in Hiroshima. The lower diagram refers to the bone marrow, the upper diagram to the colon.

The solid lines show the relations according to DS86. The ratio for Nagasaki runs parallel to that for Hiroshima but is lower by a factor of 3. The dotted lines

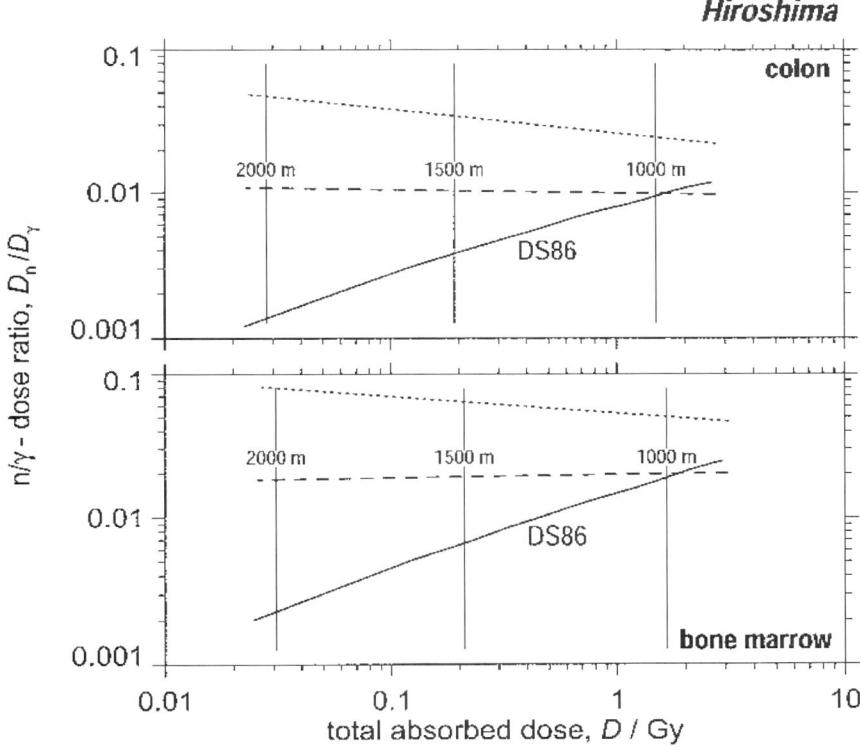

FIGURE 7-1 Ratio of neutron absorbed dose to gamma-ray absorbed dose in Hiroshima versus total dose. Solid curves correspond to the current dosimetry system, DS86; dotted curves to neutron doses increased in line with thermal-neutron activation data (Straume and others 1992); broken curves to intermediate adjustment that might be consistent with preliminary [63]Ni measurements (Chapter 3).

give the relations originally proposed by Straume and others (1992) to account for the discrepancies between thermal-neutron activation measurements and the DS86 computations. In view of the DS86 computations and the present assessment of the activation data discussed in Chapter 3, including preliminary ^{63}Ni data, this relation can now essentially be discounted. The broken lines represent a tentative, smaller modification that may not be inconsistent with the current evaluation of activation.

In analyses of mortality from or incidence of all solid cancers combined, it has been usual to refer to the colon dose, that is the dose to the deepest, most highly shielded organ. For the gamma rays, the choice is not critical; organ doses for the solid-tumor sites (averaged in terms of the ICRP tissue-weighting factors) are only about 5–10% higher than the colon dose. For neutrons, the reference to the colon is unsatisfactory because averaging over all organs at risk results in a neutron absorbed dose that is roughly 1.9 times the neutron absorbed dose to the colon. The analysis that refers the neutron dose to the colon is thus biased toward a low predicted neutron contribution to observed health effects. The neutron dose to an average organ at risk for solid cancer is close to that in the bone marrow, and later considerations will therefore use data on bone marrow as an approximation that is also adequate with regard to all solid cancers combined.

EFFECT OF THE NEUTRON CONTRIBUTION
AS INFERRED FROM RBE

Figure 7-2 shows a nonparametric representation of the excess relative risk (ERR) for solid-cancer mortality versus absorbed dose to the bone marrow in Hiroshima (adapted from Chomentowski and others 2000). At low doses, statistical imprecision makes it difficult to give a reliable value of the ERR; at doses close to 2 Gy, one recognizes some bending over of the curve that complicates any extrapolation to low doses. It is therefore reasonable to consider the total effect and the fractional effect due to neutrons at an intermediate total dose, which is chosen here to be 1 Gy.

With the neutron/gamma dose ratio 0.0075 for the colon at 1 Gy total dose (see Figure 7-1) the weighted dose at 1 Gy is 1.07 Gy (derived from 0.9925 + (10 × 0.0075)), therefore the effect contributed by neutrons is about 7% (0.075/1.07). This low value confirms the common judgment, although it depends on two assumptions; first, the choice of 10 for the relative biological effectiveness (RBE) of neutrons compared to gamma rays, which some consider low, and second, the reference site (the colon), since it underestimates the dose contributions of the neutrons.

Although the relatively high dose of 1 Gy might seem to be in line with fairly low values of the neutron RBE, it must be recognized that in DS86, at a 1 Gy total dose to the bone marrow in Hiroshima, the neutron dose is only about 15 mGy. That is in the lower range of neutron doses at which excess tumor incidence has

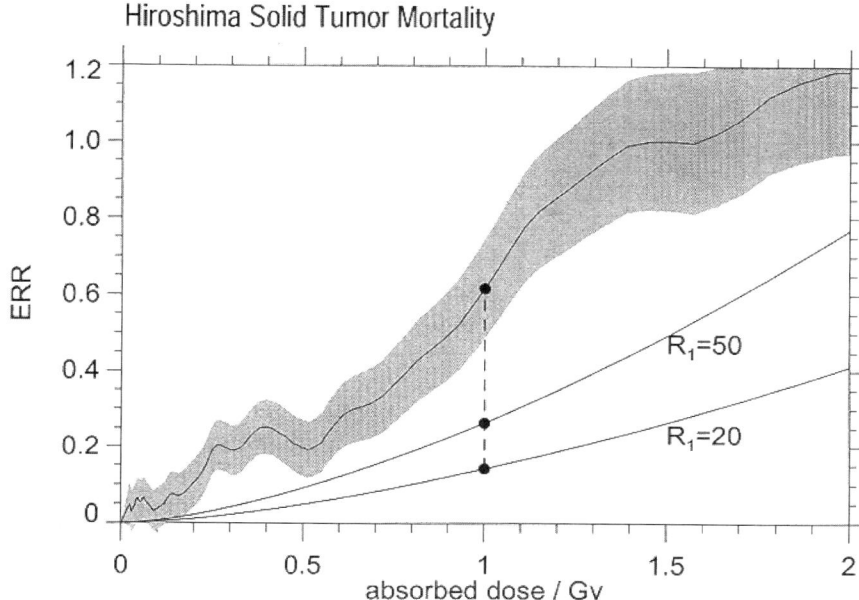

FIGURE 7-2 Nonparametric representation of the excess relative risk for solid tumor mortality in Hiroshima as a function of absorbed dose to the bone marrow (adapted from Chomentowski and others 2000). The gray shaded band indicates the standard error. The two lower curves show the effect contribution by fast neutrons that is inferred in terms of the relative biological effectiveness, R_1, of neutrons against a gamma-ray dose of 1 Gy. The effect of the neutrons is proportional to the neutron dose, but due to increasing neutron/ gamma dose ratio it increases more than linearly with total absorbed dose.

been determined in animal studies. Likewise, 1 Gy is typically near the lowest gamma-ray dose (0.5 Gy is about a minimum) at which an excess of solid tumors can be adequately measured in animal experiments. At 15 mGy (i.e., about 1 Gy gamma dose) the observed values of RBE tend to be fairly large. Results of animal studies vary, but 20 appears to be a lower value—inferred from life-shortening in mice (Carnes and others 1989; Covelli and others 1989), which has been used as a proxy for tumor incidence—and 50, consistently seen in a large series of experiments on tumor induction in rats (Lafuma and others 1989; Wolf and others 2000), appears to be a reasonable high value.

The lower RBE value of 20 implies that 50 mGy of neutrons has the same effect, E_1, as 1 Gy of gamma rays. Because of the linear dose dependence for neutrons, the 15 mGy of neutrons will contribute 0.3 E_1 to the observed effect $E_{obs} = (0.985+0.3) E_1 = 1.285 E_1$ at 1 Gy total absorbed dose. Thus, the neutrons contribute 23% of the ERR observed at a 1 Gy total absorbed dose in Hiroshima, and

77% is due to gamma rays. Likewise, one obtains with the upper RBE value of 50 against 1 Gy of gamma rays a neutron-effect contribution of $0.75 E_1$, of a total effect of $E_{obs} = 0.985 + 0.75 E_1 = 1.735 E_1$; that is, the neutrons contribute 43% and the gamma rays contribute 57% of the effect. The major point is that regardless of where in the interval 20–50 the true RBE lies, there is a substantial contribution of neutrons (as indicated by DS86) to the effects at about a 1 Gy total absorbed dose in Hiroshima.

COMPARISON OF NEUTRON EFFECTS
IN HIROSHIMA AND NAGASAKI

If the same calculations are made for Nagasaki with only one-third as great a neutron contribution as that in Hiroshima—5 mGy instead of 15 mGy at a 1 Gy total absorbed dose—then the calculations can be summarized as in the following table (Table 7-1).

The results in Table 7-1 show that from a consideration of reasonable RBEs for neutrons, the effects to be expected for the same total absorbed dose of 1 Gy should be 18–40% higher in Hiroshima than in Nagasaki only because of the greater number of neutrons in Hiroshima, as calculated in DS86. Obviously, as noted below, increasing the neutron contribution in Hiroshima because of thermal-neutron (and fast-neutron) activation measurements would increase the ratio of effects at Hiroshima versus Nagasaki further. In fact, it is well known that intercity differences, seem to exist (Pierce and others 1996), with ERR estimates for Hiroshima 1.5–2.0 times as great as those for Nagasaki and excess absolute risk estimates 1.2-1.5 times as great as those for Nagasaki, on the basis of cancer-mortality data from 1950-1990. The complexity of the issues involved in determining intercity differences, especially a bias that might have led to the large ratio of ERR estimates, are described in some detail in Pierce and others (1996) and Pierce and Preston (2000).

None of those observations leads to clearly significant differences. Although the intercity difference is evidently in the direction of greater neutron effects in

TABLE 7-1 Neutron/Gamma Dose Calculations for Hiroshima and Nagasaki

Total Dose	Gamma-Ray Dose	Neutron Dose	Neutron RBE	Total Weighted Dose	Hiroshima/Nagasaki Effect Ratio for RBE
			Hiroshima		
1 Gy	0.985 Gy	0.015 Gy	10	1.135 Gy	1.09
1 Gy	0.985 Gy	0.015 Gy	20	1.285 Gy	1.18
1 Gy	0.985 Gy	0.015 Gy	50	1.735 Gy	1.40
			Nagasaki		
1 Gy	0.995 Gy	0.005 Gy	10	1.045 Gy	
1 Gy	0.995 Gy	0.005 Gy	20	1.095 Gy	
1 Gy	0.995 Gy	0.005 Gy	50	1.245 Gy	

Hiroshima, it is impossible with the current information to ascribe differences in effects to the neutrons in DS86 alone and therefore to determine neutron RBEs from them. It is useful to note, however, that a neutron RBE of 50 would explain, with DS86 unmodified, the reported intercity difference.

Taking the "speculation" in this illustration a step further, we can also note that the risk of all solid cancer and leukemia combined, 12% Sv^{-1} (UNSCEAR 2000), has been derived from the atomic-bomb survivor data with an assumed RBE of 10 and with reference to the colon dose. That corresponds to the assumption of an RBE of 5 with reference to the average organ dose and with an average 11 mGy of neutrons at a 1 Gy total dose in the combined sample; the effect at 1 Gy was taken to be $E_{obs} = [0.989 + 5 (0.011)] E_1 = 1.044 E_1$. Using an RBE of 50, one would have $E_{obs} = [0.989 + 50(0.011)] E_1 = 1.539 E_1$. The effect, E_1, of 1 Gy of gamma rays—that is the risk estimate—is thus reduced by the factor 1.044/1.539, that is, it is 8.1% Sv^{-1} (this will be applicable directly to acute gamma rays, although in ICRP-NCRP procedures it would be divided by a DDREF of 2 to become 4.1% for the risk at low dose rates, slightly less than the accepted nominal value of 5.0% Sv^{-1} but well within the range of uncertainty for this estimate).

IMPLICATIONS OF THE NEUTRON DISCREPANCY

The potential implications of the neutron discrepancy—the postulate that thermal activation implies a substantially larger fast neutron component in Hiroshima than indicated by DS86 but the same fast neutron component in Nagasaki as indicated by DS86—can first be exemplified by a consideration of the large modification of neutron doses that was thought to be in line with the trend of the thermal-neutron measurements during the last few years. With the full modification of Straume and others 1992 (see dotted line in Figure 7-1) the neutron/gamma ray dose ratio at a 1 Gy total dose would be 0.055. In analogy to the above calculations, that would imply a neutron-effect contribution of 54% at a 1 Gy total dose if the RBE were 20 and a contribution of 74% if it were 50. Although the full modification of Straume and others now can probably be ruled out because of our evaluation of the recent fast-neutron measurements (see Chapter 3) with ^{63}Ni, the example shows why the resolution of the discrepancy is important to appreciate how the risk must be apportioned between neutrons and gamma rays.

The activation measurements that have already been performed provide general confirmation of DS86, at distances around 1000 m, corresponding to total doses close to 2 Gy. At 1 Gy an increase of the neutron dose—perhaps from 15 mGy to 20 mGy—is still possible, but a larger increase seems unlikely. However it is clear that any additions to the neutron component at Hiroshima will suggest a higher effect contribution by the neutrons.

At smaller doses, a more substantial increase in the neutron/gamma ray dose ratio, possibly of 3–5 at 0.2 Gy, cannot be excluded and might indeed be the final outcome of fast-neutron measurements. That could have some impact on the risk coefficient primarily for gamma rays (Kellerer and Walsh 2001), although uncertainties will inevitably be very large. There is some conflict between the general experience, in radiobiological studies, of upward curvature in the dose-effect relations and the finding that the dose dependence for solid tumors in Hiroshima and Nagasaki is seemingly linear. More curvature is, of course, indicative of a lower risk coefficient for gamma rays. Because more neutrons at low doses in Hiroshima would explain at least part of the seeming linearity of the overall dose dependence, it would also be consistent with a somewhat lower gamma-ray coefficient. In any case both the gamma-ray risk estimates and the neutron-risk estimates will depend, but probably not critically, on the successful resolution of the neutron discrepancy.

In conclusion, it is probably already clear from the preliminary ^{63}Ni measurements that the neutron discrepancy is smaller than at first thought and possibly within the range of uncertainties in the contribution of neutrons in Hiroshima and Nagasaki given by DS86. It is clear, however, that according to the illustration given here the effects of these neutrons, even at the DS86 level, are not negligible and, when allowed for, tend to lower the gamma-ray risk estimates slightly.

8

Conclusions and Recommendations

DS86 is clearly a much more complete and sophisticated dosimetry system than any of its predecessors, and it calculates the organ doses in the survivors with considerable accuracy. The main component of the dose in organs deep in the body is gamma radiation, and measurement of this component with thermoluminescent techniques yields excellent agreement with the calculations of DS86 to within the accuracy of the measurement technique, possibly about ±10% and in any case within the range of expected uncertainty. However, the agreement might be improved further if the energy response of the TL measurements were reexamined. The most significant of the factors that potentially affect the TL measurements is the increase in the response of the TL with low gamma-ray energy. An estimate of the correction needed to account for the energy response is a decrease in the reported measurement value by an arbitrary 20%.

The neutron component of the dose is small, especially in Nagasaki, and considerably less certain than the gamma-ray component at the time of writing of this report. In 1986, it was apparent that unresolved discrepancies existed between measurements of thermal-neutron activation in cobalt and in europium and calculations of neutron fluence in Hiroshima and Nagasaki. In the revisions proposed since DS86, the discrepancies have essentially been resolved for Nagasaki (mainly because of new and finer group calculations and cross-section improvements) but have tended to become worse for Hiroshima with the addition of ^{36}Cl thermal-neutron activation measurement techniques, possibly amounting to thermal-neutron fluence 1500 m from the epicenter greater by a factor of almost 10 than calculated in DS86.

Determined efforts in the last few years to establish the magnitude of the "neutron problem" and to explain it by examining critically the uncertainties in the measurements themselves have not resulted in definitive conclusions. Issues related to the background and other uncertainties of some of the samples need to be explored further. And possible modifications of the spectrum issuing from the Hiroshima weapon has not produced a new source that is consistent with the reported measurements of thermal activation at great distances or satisfied the dynamics of the bomb explosion and the sulfur-activation measurements of fast neutrons made close to the bomb soon after its explosion.

This committee and others still working on these problems recognized that measurements of the fast-neutron component were essential to resolve this problem. A new method—measuring ^{63}Ni produced by fast neutrons in copper samples (^{63}Cu (n,p) ^{63}Ni)—was proposed and strongly endorsed by this committee, which urged immediate support for such investigations (NRC 1996). Suitable samples of copper irradiated at the time of the bombing in Hiroshima (August 6, 1945) have been obtained and are being measured in Japan (with ^{63}Ni radioactivity; T½ 100 y) and in the United States and Germany (by accelerator mass spectrometry). More samples at strategic locations are still urgently being sought in Hiroshima and Nagasaki to expand the measurement program and make it as complete as possible.

At 1000 m, it appears that DS86 is not significantly in disagreement with measurements. Even at greater distances, the discrepancy is probably less than previously reported. However, the data are not certain enough to allow a good estimate of the discrepancy at great distances. Further measurements of ^{63}Ni and ^{36}Cl and improved uncertainty analyses are needed for better definition of the magnitude of the disagreement in the neutron fluence and confirmation of the ^{32}S data for Hiroshima. The new measurements should use pre-established data-quality objectives. Some of the previously reported data, in particular the ^{36}Cl data, need to be reanalyzed to resolve possible errors in the original reported results. Future measurements should include secondary-reference standard reagents and analytical blanks, and when it is appropriate, an isotopic tracer should be processed with each field sample. Preliminary results of the ^{63}Ni fast-neutron measurements suggest that the discrepancy in Hiroshima is smaller (perhaps a factor of 3-5, not a factor of 10, at 1500 m) than suggested by the thermal-neutron activation measurements. Reconciling the relaxation lengths derived from some of these measurements is difficult. Further exploration of the apparent neutron discrepancy will continue with as many additional samples as feasible.

If it turns out that the discrepancy is indeed small and perhaps within the uncertainties of calculation and measurement, the doses to survivors can be regarded as reliable within a given uncertainty range. For the estimation of gamma-ray risk, it has been noted in this report that the neutron component—although small, with reasonable values of RBE in the range of 20–50—could contribute appreciably to the total effect even without any increase in neutron fluence in Hiroshima. A somewhat greater neutron contribution will result if neutron fluence is eventually increased at great distances there. However, it is clear again that although the risk from gamma-rays might, as a consequence, be a little less than previously estimated, it is well within the range of uncertainty known to exist for these estimates (NCRP 1997).

SUMMARY

Although DS86 is a good system for specifying dose to the survivors and for assessing risk, it needs to be updated and revised. Uncertainties have not been fully evaluated and might amount to more than the 25-40% in fractional standard deviations of parameters (Kaul and Egbert 1989). A number of parameters in DS86 have been improved and should be implemented. While the calculated gamma-ray fluences agree well with values measured with thermoluminescence and constitute the main component of the dose to the survivors, more work needs to be done to establish the magnitude of the neutron component and to assess the extent to which the neutron component (small in DS86) affects (lowers) the estimates of gamma-ray risk.

The committee offers the following recommendations regarding the revision of DS86 that is clearly needed and that hopefully will be completed in 2002:

- The present program of [63]Ni measurements should be pursued to completion.
- All thermal-neutron activation measurements, particularly those with [36]Cl and [152]Eu, should be reevaluated with regard to uncertainties and systematic errors, especially background (see Chapter 3).
- Critical efforts to understand the full releases from the Hiroshima bomb by Monte Carlo methods should be continued.
- Adjoint methods of calculation (i.e., going back from the field situation to the source term) should be pursued to see whether they help solve the neutron problem.
- Local shielding and local-terrain problems should be resolved.
- The various parameters of the Hiroshima explosion available for adjustment, including the height of burst and yield, should be reconsidered in the light of all current evidence in order to make the revised system as complete as possible.
- A complete evaluation of uncertainty in all stages of the revised dosimetry system should be undertaken and become an integral part of the new system.
- The impact of the neutron contribution on gamma-ray risk estimates in the new system should be determined.

Appendix A

The RERF Dosimetry Measurements Database and Data Collection for the Dosimetry Reassessment

The RERF Dosimetry Measurements Database attempts to compile a detailed list of analytically useful data on measurements of thermoluminescence and neutron activation in Hiroshima and Nagasaki. The database is intended to contain an entry for every such measurement that has been made and documented. The database is prepared in Access™ and has custom screens for data entry and checking that are programmed in Visual Basic™, as shown in Figures A-1 and A-2. Source documents for the data are detailed in a table of references (Table A-2). Chapters and appendixes of the DS86 final report are listed separately in the table of references (Table A-2). In addition to the DS86 final report, the database includes listings for 42 published papers and various other reports, proceedings, and notes from meetings.

In addition to the table of references, there are also linked tables of samples, subsamples, and measurements in the database. Data are entered exactly as they appear in source documents, sometimes with extensive annotation in the notes field to aid in interpretation. RERF maintains and updates the database by vetting the existing entries and adding new data. A summary cross-tabulation of measured samples in the database is given in Table A-1 for samples that fit the "surface, line of sight" criterion discussed in Appendix B. The database also contains results for a number of other samples, notably core samples that yield information at various depths in rock or concrete.

Initially, efforts were made to maximize the information in the database from available source documents. RERF supported and participated in the efforts of the Committee on Dosimetry for the RERF to obtain detailed information directly from investigators for the purpose of a comprehensive uncertainty analysis. In November 1998, Mr. Lowder and Dr. Takashi Maruyama, accompanied by Dr. Cullings, visited

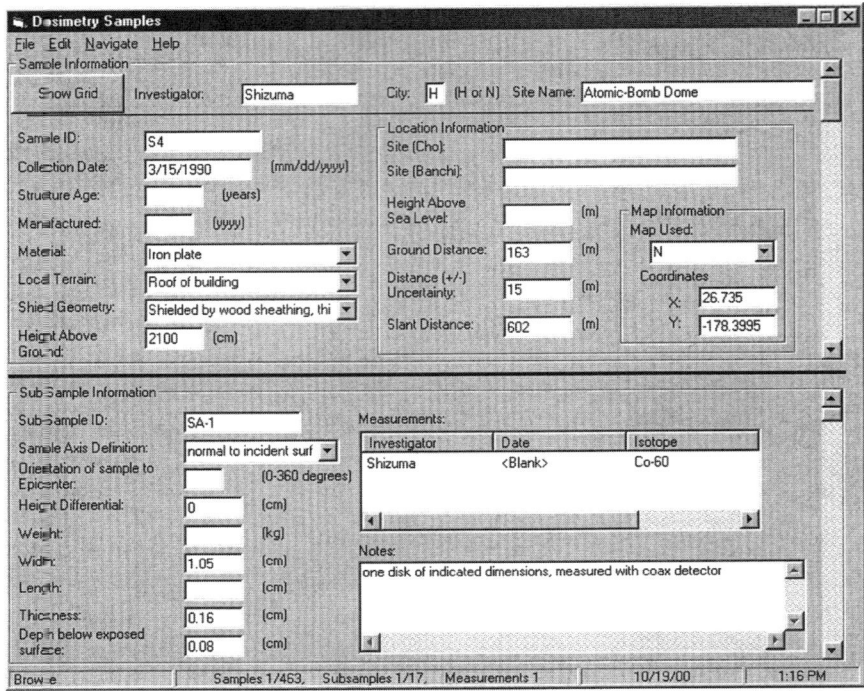

FIGURE A-1 Sample data-entry screen of RERF dosimetry measurement database.

the laboratories at Hiroshima University Geniken (Dr. Hoshi), Hiroshima University Saijou Campus (Dr. Shizuma), Kanazawa University (Dr. Nakanishi), and N.I.R.S. at Chiba near Tokyo (Maruyama and Kumamoto) for meetings, tours, and consultations. They also met with Dr. Fujita and Mr. Watanabe, of RERF, who have extensive personal knowledge and records of sample collection, storage, and distribution.

In December 1998, a detailed questionnaire prepared by Mr. Lowder, which is included at the end of this appendix, was sent to all investigators who had made measurements of interest for the uncertainty analysis. Written responses were received only from Dr. Hamada and Dr. Kato.

In December 1999, Dr. Maruyama and Dr. Cullings again visited the laboratory of Dr. Nakanishi with a detailed list of requested information.

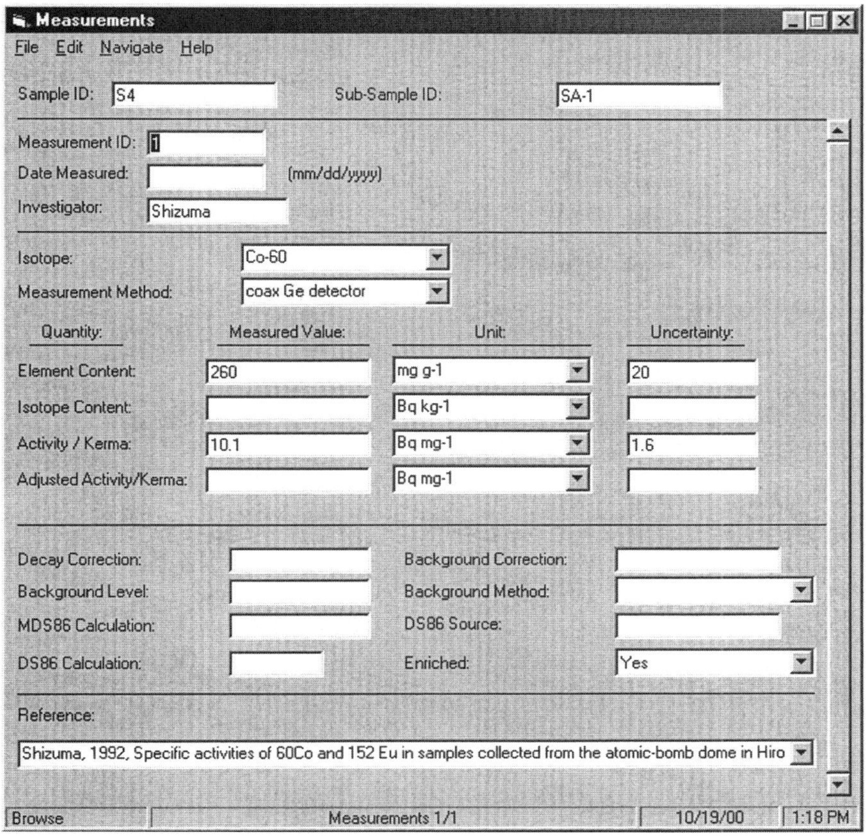

FIGURE A-2 Measurements data-entry screen of RERF dosimetry measurement database.

Dr. Cullings has exchanged letters and materials with Dr. Shizuma. In April 2000, Dr. Cullings wrote letters to Dr. Nakanishi, Dr. Iimoto, Dr. Hamada, Dr. Shimazaki (Dr. Okumura's group in Nagasaki) and Dr. Shizuma, requesting spectra and additional detailed information on background issues. Responses were received from Dr. Iimoto, Dr. Kimura (Dr. Hamada's group), and Dr. Shizuma.

In September 2000, Dr. Cullings attended a meeting of the Hiroshima dosimetry group and made a brief presentation regarding environmental background samples and questions concerning the accuracy of the trapezoidal approximation of background versus peak counts at low sample radioactivity levels.

The information obtained from all those inquiries has been archived at RERF and has been used to augment and correct the database as appropriate.

TABLE A-1 Measured Locations in the RERF Database[a] (by Site Name, Ground Range in m, and Azimuth in Degree)

Hiroshima ^{32}P (measured in 1945)

Site	Range	Azimuth		Site	Range	Azimuth
?	76	242		?	660	?
?	178	4		?	682	75
?	305	79		?	705	?
?	331	?		?	721	296
?	331	?		?	968	?
?	358	311		?	1025	299
?	417	56		?	1080	?
?	433	132		?	1305	?
?	470	?				

Hiroshima ^{36}Cl

Site	Range	Azimuth		Site	Range	Azimuth
Saikouji	94	265		Hiroshima University E Building[b]	1354	165
Motoyasu Bridge	102	245		Teishin Hospital (Communications Hosp.)	1368	44
Aioi Bridge[b]	300	318		Hiroshima University Radioisotope Bldg[b]	1427	163
Fukoku Seimei Building[b]	317	131		Hiroshima University Seifu Dormitory[b]	1427	176
Gokoku Shrine[b]	398	351		Red Cross Hospital North Bldg[b]	1469	182
Kirin Beer Hall[b]	664	112		Red Cross Hospital[b]	1501	180
Chugoku Electric Power Co.	676	175		Hiroshima Postal Savings Bureau	1606	177
Sinkoji[b]	870	325		Hiroshima Bank of Credit[b]	1703	350
Old NHK Building[b]	988	83		Hiroshima Commercial High School[b]	2863	222
Hiroshima City Hall	1000	177				
Ganjioji[b]	1029	32				
Tokueiji[b]	1140	108				
Jyunkyoji[b]	1217	122				
Hosenji[b]	1225	334				
Hiroshima University Elementary School[b]	1269	167				

Hiroshima ^{60}Co

Site	Range	Azimuth		Site	Range	Azimuth
Shima Hospital	0	0		Sentry Box	640	131
Hiroshima Post Office	55	313		Kirin Beer Hall	670	113
Saikou-ji	94	302		Chugoku Electric Power Co.	687	174
Motoyasu Bridge	128	248		Kodokan	720	274
Atomic-Bomb Dome	163	~308?		Water Trough	793	351
Monument of victory	213	50		Hiroshima City Hall	1014	180
Sumitomo Bank	250	?		Powder Magazine	1197	11
Yasuda Seimei Building	257	116		Yokogawa Bridge	1295	343
Hiroshima Bank	269	206		Yokogawa Bridge	1295	343
Aioi Bridge	300	325		Red Cross Hospital	1481	180
Fukoku Seimei Building	331	146		Red Cross Hospital	1484	180
Honkawa Primary School	373	297		Hiroshima Bank of Credit	1703	350
Fukuromachi Primary School	441	?				

TABLE A-1 (*Continued*)

Hiroshima [152]Eu					
Shima Hospital	0	0	Shirakami Shrine G1	478	163
Sei Hospital	55	334	Naka Denwa-Kyoku		
Hiroshima Post Office	55	313	(Telephone Ofc)	529	131
Saikou-ji	94	302	Honkawa stone lantern		
Motoyasu Bridge, Pillar 2	101	249	(gangi)	531	250
Saikou-ji, grave	107	265	Seiju-ji	546	305
Daiichi Bank	129	118	Kyoden-ji	548	272
Motoyasu Bridge Pillar 1	132	248	Sanin Godo Bank	618	96
Chiyoda Seimei Bldg	132	84	Kakomachi stone wall	629	232
Atomic-Bomb Dome	137	308	Sorazaya Shrine	653	326
Motoyasu Bridge, Railing	146	243	Myocho-ji	654	288
Atomic-Bomb Dome	150	~308	Chugoku Electric Power Co.	655	174
Atomic-Bomb Dome	161	307	Akisaya-cho	675	35
Nihon Seimei Bldg	163	147	Hiroshima Castle	694	32
Atomic-Bomb Dome	165	~308	Kawaramachi stone wall	714	238
Atomic-Bomb Dome	168	~308	Chugoku Electric Power Co.	720	174?
Atomic-Bomb Dome	173	~308	Kodo Primary School	720	274
Banker's Association	178	155	Choukaku-ji	849	76
Rest House	189	~90	Tamino's House	875	125
Monument of victory	213	50	Hiroshima Prefectural Office	881	217
Sensho-ji	229	148	Honkei-ji	893	186
Hiroshima Bank			Enryu-ji	912	0
(Geibi Bank)	250	89	Yorozuyo Bridge stone wall	924	209
Hiroshima Bank	250	89	Shingyo-ji	927	329
Yasuda Seimei Bldg	253	116	Teramachi stone wall	949	334
Sumitomo Bank	255	101	Hiroshima Radio Station	988	83
Aioi Bridge P1	258	318	Hiroshima City Hall	1017	181
Daido Seimei Bldg	269	87	Tenma bashi[b]	1029	275
Jisen-ji	272	266	Kozen-ji	1163	123
Fukoku Seimei Bldg	317	146	Iwamiya-cho	1197	96
Honkawa stone wall	344	261	Hiroshima University	1255	~165
Gogoku Shrine, Lantern	344	351	Hiroshima University	1274	~165
Honkawa Primary School	359	286	Hiroshima University	1298	~165
Gokoku Shrine, Marble	377	347	Sumiyoshi shrine[b]	1307	213
Gokoku Shrine, Guarding Lion	381	349	Hiroshima University	1328	~165
Honkawa Primary School	389	360	Hiroshima University,		
Gokoku Shrine	398	351	Primary School	1335	166
Sanyo Memorial Hall	410	150	Kyo Bridge, Railing	1357	91
Motomachi Stone Wall 1	420	336	Teishin Hospital		
Seigen-ji	427	141	(Communications Hospital)	1370	44
Honkawa Bridge stone wall	434	261	Kannon bashi[b]	1618	237
Geibi Bank, Tsukamoto			Hiroshima Commercial		
Branch	465	261	High School[b]	2863	222

(*continued*)

TABLE A-1 Measured Locations in the RERF Database[a] (by Site Name, Ground Range in m, and Azimuth in Degree) (*Continued*)

Hiroshima ⁶³Ni

Site	Range	Azimuth
Atomic-Bomb Dome	163	~308?
Bank of Japan[b]	367	153
Soy Sauce Brewery[b]	948	90
City Hall[b]	1013	180
Univ. Elementary School[b]	1304	167
Hiroshima University Radioisotope Bldg[b]	1461	163
Sumitomo Bank[b]	1880	88

Hiroshima TLD

Site	Range	Azimuth
Shima Hospital ?	14	346
Motoyasu bashi ?	113	258
?	188	330
?	194	50
"Zaimoku-cho, Dempuku-ji"	400	SW
"Zaimoku-cho, Seigan-ji"	420	SW
"Zaimoku-cho, ?"	430	SW
?	460	117
Naka Telephone Office	507	129
Naka Telephone Office	523	132
Sanin Bank	621	95
?	623	89
Choguku Electric Co.	665	174
Choguku Electric Co.	692	175
?	715	92
"Ninomaru, Hiroshima Castle"	750	NNE
Nishishin-machi	800	W
"Nishishin-machi, koen-ji"	960	NNW
"Nishishin-machi, Shozen-ji"	970	NNW
"Honmaru, Hiroshima Castle"	980	NNE
Nobori-cho (Japanese house)	1131	85
HUPS	1271	167
HUPS	1282	168
HUPS	1298	167
HUPS	1316	167
HUPS	1338	166
HUFS-I	1338	168
HUFS	1377	165
HUS	1378	168
HUFS	1387	167
HUFS	1388	166
HUFS-E	1388	169
HUFS	1393	166
HUFS	1397	166
HUFS	1401	167
HUFS	1422	166
HUFS	1425	167
HUFS	1426	165
HUFS	1428	167
HUFS	1428	166
HUFS	1433	167
HUFS	1449	165
HUFS	1450	167
HUFS	1451	165
Red Cross Hospital	1452	206
Red Cross Hospital	1452	181
HUFS	1457	167
HUFS	1459	166
HUFS	1460	166
HUFS	1461	167
HU Radioisotope Bldg.	1462	163
Red Cross Hospital	1501	180
Postal Savings Bureau	1591	177
Chokin-kyoku (Postal Savings)	1597	177
Postal Savings Bureau	1604	178
Postal Savings Bureau	1605	177
Chokin Kyoku (Postal Savings)	1613	178
Postal Savings Bureau	1613	177
Postal Savings Bureau	1631	176
Japan Elec. Meters Insp. Corp.	1793	356
"Meisen-ji" "Oni-gawara"	1909	107
HUT (HUFE)	2051	178
"Hiramoto" "Oni-gawara"	2053	253
HUFE	2054	180
Kirihara house	2453	287
Ryomatsu-sho (Provisions Depot)	3133	168

TABLE A-1 (*Continued*)

			Nagasaki ^{36}Cl		
Nagasaki University			Fuchi Middle School	1156	203
Hospital	650	144	Konpira-san Anti-aircraft		
Mitsubishi Steel	1075	181	Battery[b]	1580	127

			Nagasaki ^{60}Co		
?	18	?	?	353	?
?	39	?	?	460	?
?	63	?	?	472	?
?	82	?	Nagasaki Medical		
?	92	?	School Building	520	125
?	93	?	Shiroyama School[b]	540	
?	96	?	?	561	?
?	118	?	Nagasaki Medical		
?	249	?	School Building	590	12
Takatani House [b]	290	?	Nagasaki University		
?	307	?	Hospital[b]	653	144
?	330	?	Motoki Bridge[b]	780	N
?	343	?	Mitsubishi Steel[b]	935	S
?	347	?	COMM SCHOOL	1030	~300?

			Nagasaki ^{152}Eu		
?	19	?	N5	427	?
Shimono-kawa	20	NNE or ESE or S	?	432	?
?	40	?	?	435	?
Shimono-kawa	48	NNE or ESE or S	?	457	?
?	62	?	Urakami Church[b]	465	60
Shimono-kawa	80	NNE or ESE or S	?	474	?
?	80	?	?	523	?
Shimono-kawa	93	NNE or ESE or S	N6	528	?
?	93	?	N7	555	?
?	94	?	?	560	?
?	96	?	?	590	?
Shimono-kawa	100	NNE or ESE or S	?	591	?
N1	100	?	?	628	?
?	109	?	?	635	?
Shimono-kawa	110	NNE or ESE or S	?	641	?
?	115	?	N8	645	?
N2	226	?	Gokoku shrine	651	303
?	247	?	Nagasaki University		
Urakami-gawa	250	WSW	Hospital[b]	653	144
Urakami-gawa	255	WSW	?	668	?
Urakami-gawa	293	WSW	?	682	
Urakami-gawa	300	WSW	Nanzan school[b]	704	22
?	308	WSW	?	710	?
Yana bashi[b]	311	296	?	751	?
?	312	?	?	776	?

(*continued*)

TABLE A-1 Measured Locations in the RERF Database[a] (by Site Name, Ground Range in m, and Azimuth in Degree) (*Continued*)

Nagasaki ¹⁵²Eu

?	313	?	?	782	?
?	329	?	?	794	?
?	342	?	Shimoda house[b]	812	149
?	346	?	?	848	?
?	349	?	N9	871	?
?	352	?	Prefectural gymnasium[b]	871	180
N3	362	?	?	916	?
N4	379	?	?	934	?
?	389	?	St. Maria school[b]	952	155
			Anakoboji Temple	1020	ESE
			Sakamoto-cho[b]	1039	153
			Ide residence	1060	W
			Maruo cho[b]	2850	186

Nagasaki TLD

?	95	?	?	970	?
Matsuyama-cho	100	ENE	Uragami-cho	980	N
Oka-machi	230	NW	?	1020	?
Yamazoto-cho	330	NE	Sakamoto Cho Cemetery	1039	153
Shiroyama Elementary School	350	W	?	1046	?
Shiroyama-cho	400	W	?	1066	154
?	520	?	Ceramic (Nishimachi)	1075	3
Urakami	521	57	?	1173	?
Shiroyama-cho	600	SW	Zenza	1426	168
?	635	?	?	1427	?
Ueno-cho	650	NE	Ieno wall	1432	355
Nagasaki University Hospital	653	178	Nagasaki University		
Brazier (Shiroyama)	730	276	Hospital Morgue	1435	167
Shiroyama-cho	740	NE	Ieno-cho roof	1564	360
Sakamoto-cho	760	SE	Yamada Oil Warehouse	2043	176
?	836	?	Inasa	2049	175
?	836	?	Inasa	2051	175
?	860	?	Inasa	2062	176
?	875	?	Chikugo	2328	156
?	935	?			

[a] As of April 10, 2001. Some of the indicated sites have measurements on multiple samples or cores for depth profiles. Measurements lacking precise azimuthal information are not shown in the maps (Plates 1 and 2). Measurements of ¹⁵⁴Eu and ⁴¹Ca are not included in the table, as they currently exist at only one or two locations. "HU" = Hiroshima University.

[b] Measurement not yet published.

TABLE A-2 References in the RERF Dosimetry Measurements Database

RefID	First Author	Year	Title	Journal
1	Egbert	1995	Computerized data acquisition and retrieval system for archival of Hiroshima and Nagasaki A-bomb activation measurements and calculations	Book (Science Applications International Corporation)
2	Gritzner	1987	Sulfur activation at Hiroshima	DS86 Vol. 2:283–292
3	Hasai	1987	^{152}Eu depth profile of a stone bridge pillar exposed to the Hiroshima atomic bomb: ^{152}Eu activities for analysis of the neutron spectrum	Health Phys. 53:227–239
4	Kato	1990	Gamma-ray measurement of ^{152}Eu produced by neutrons from the Hiroshima atomic bomb and evaluation of neutron fluence	Jpn. J. Appl. Phys. 29:1546–1549
5	Kato	1990	Accelerator mass spectrometry of ^{36}Cl produced by neutrons from the Hiroshima bomb	Int. J. Radiat. Biol. 58:661–672
6	Kaul	1987	Calculation of dose in quartz for comparison with thermoluminescence dosimetry measurements	DS86 Vol. 2:204–241 (Appendix 11 to Chapter 4)
7	Kerr	1983	Tissue kerma vs distance relationships for initial nuclear radiation from the atomic bombs Hiroshima and Nagasaki	First 1983 RERF Workshop: 57–103
8	Kerr	1990	Activation of cobalt by neutrons from the hiroshima bomb	ORNL 6590
9	Milton	1968	Tentative 1965 radiation dose estimation for atomic bomb survivors	ABCC Technical Report 1–68
10	Nakanishi	1987	Residual neutron-induced radio-activities in samples exposed in Hiroshima	DS86 Vol. 2:310–319
11	Nakanishi	1991	Residual neutron-induced radio-nuclides in samples exposed to the nuclear explosion over Hiroshima: Comparison of the measured values with the calculated values	J. Radiat. Res. S:69–82
12	Nakanishi	1993	Calculated and measured ^{152}Eu activity in roof tiles exposed to atomic bomb radiation in Nagasaki (in Japanese)	1992 research report on effects of the atomic bombs

(*continued*)

TABLE A-2 References in the RERF Dosimetry Measurements Database
(*Continued*)

RefID	First Author	Year	Title	Journal
13	Okumura	1997	Reassessment of Atomic bomb neutron doses (in Japanese)	FY 1996 Report of Research Group on Atomic Bomb Related Symptoms
14	Roesch	1987	Book (US-Japan joint reassessment of atomic bomb radiation dosimetry in Hiroshima and Nagasaki)	
15	Shibata	1994	A method to estimate the fast-neutron fluence for the Hiroshima atomic bomb	J. Phys. Soc. Jpn. 63:3546–3547
16	Shizuma	1992	Specific activities of ^{60}Co and ^{152}Eu in samples collected from the atomic-bomb dome in Hiroshima	J. Radiat. Res. 33:151–162
17	Shizuma	1992	Low-background shielding of Ge detectors for the measurement of residual ^{152}Eu radioactivity induced by neutrons from the Hiroshima atomic bomb	Nuclear Instruments and Methods in Physics Research B66:459–464
18	Shizuma	1993	Residual ^{152}Eu and ^{60}Co activities induced by neutrons from the Hiroshima atomic bomb	Health Phys. 65:272–282
19	Shizuma	1997		Notes from October 1997 meeting
20	Straume	1992	Neutron discrepancies in the DS86 Hiroshima dosimetry system	Health Phys. 63:421–426
21	Straume	1994	Neutrons confirmed in Nagasaki and at the army pulsed radiation facility: Implications for Hiroshima	Radiat. Res. 138:193–200
22	Tatsumi-Miyajima, J	1991	Physical dosimetry at Nagasaki— ^{152}Eu of stone embankment and electron spin resonance of teeth from atomic bomb survivors	J. Radiat. Res. Suppl.: 83–98
23	Hashizume	1967	Estimation of the air dose from the atomic bombs in Hiroshima and Nagasaki	Health Phys. 13:149–161
24	Hashizume	1983	Present plans for dose reassessment experiments by the Japanese	Second 1983 RERF Workshop: 7–12
25	Nakanishi	1983	^{152}Eu in samples exposed to the nuclear explosions at Hiroshima and Nagasaki	Nature 302:132–134

TABLE A-2 (*Continued*)

RefID	First Author	Year	Title	Journal
26	Maruyama	1987	Comments on ^{60}Co measurements	DS86 Vol. 2:335–339 (Appendix 16 to Chapter 5)
27	Sakanoue	1987	In situ measurement and depth profile of residual ^{152}Eu activity induced by neutrons from the atomic bomb in Hiroshima	DS86 Vol. 2:261–265 (Appendix 7 to Chapter 5)
28	Hoshi	1989	^{152}Eu activity induced by Hiroshima atomic bomb neutrons: Comparison with the ^{32}P, ^{60}Co, and ^{152}Eu activities in dosimetry system 1986	Health Phys. 57:831–837
29	Kimura	1990	Determination of specific activity of cobalt (^{60}Co/Co) in steel samples exposed to the atomic bomb in Hiroshima	J. Radiat. Res. 31:207–213
30	Saito	1987	Radiochemical estimation of neutron fluence of Hiroshima and Nagasaki atomic bombs	DS86 Vol. 2:249–251 (Appendix 4 to Chapter 5)
31	Hoshi	1987	Data on neutrons in Hiroshima	DS86 Vol. 2:252–255 (Appendix 5 to Chapter 5)
32	Straume	1995	Personal communication (SAIC DB)	Personal communication (SAIC DB)
33	Straume	1997	ABCC-RERF 50th Anniversary	ABCC-RERF 50th Anniversary
34	Hoshi	1985	Distribution of ^{152}Eu in bridge	Summary reports of grants in aid for Monbusho 1985 pp 17–19
35	Nakanishi	1986	DS86 (SAIC DB)	DS86 (SAIC DB)
36	Nakanishi	1986	86-report (SAIC DB)	86-report (SAIC DB)
38	Egbert	1997	SAIC database	SAIC database
39	Hashizume	1967	Estimation of air dose from the atomic bombs, Hiroshima and Nagasaki	ABCC TR 6–67
40	Loewe	1981	Revised estimates of neutron and gamma-ray doses at Hiroshima and Nagasaki	Germantown Conference Proceedings: 25–51
41	Kerr	1981	Findings of a recent Oak Ridge National Laboratory review of dosimetry for the Japanese atom-bomb survivors	Germantown Conference Proceedings: 52–97

(*continued*)

TABLE A-2 References in the RERF Dosimetry Measurements Database (*Continued*)

RefID	First Author	Year	Title	Journal
42	Maruyama	1981	Dosimetry studies in Japan	Germantown Conference Proceedings: 201–208
43	Kato	1982	Aioi Bridge	Proc. Hiroshima University of Geniken 23:179–186
44	Hamada	1983	Measurement of 32p activity induced in sulfur in Hiroshima	First 1983 RERF Workshop: 45–56
45	Loewe	1983	Calculation and interpretation of in situ measurements of initial radiations at Hiroshima and Nagasaki	First 1983 RERF Workshop: 138–155
46	Okajima	1983	Measurement of neutron-induced ^{152}Eu radioactivity in Nagasaki	First 1983 RERF Workshop: 156–168
47	Hamada	1983	^{32}P activity induced in sulfur in Hiroshima: reevaluation of data by Yamasaki and Sugimoto	Second 1983 RERF Workshop: 52–55
48	Pace	1983	Sulfur activation in electric pole insulators in Hiroshima	Second 1983 RERF Workshop: 56–58
49	Sinclair	1983	Rapporteur's report	Second 1983 RERF Workshop: 59–63
50	Kato	1984	Aioi Bridge	Hiroshima Igaku 37:345–348
51	Maruyama	1985	Commentary on ^{60}Co measurements	Unpublished draft (SAIC)
52	Nakanishi	1985		Monbusho report pp 25–43
53	Kato	1985		Monbusho report pp 44–52
55	Hashizume Tajima	1985	Concerning rebar ^{60}Co rebar measurements	Letter to Dean Kaul
56	Kerr	1985	ORNL iron surface measurements (not to appear in publications for record only)	Memo to Joe Pace
57	Okajima	1985	Draft report (alternate version appears in green book)	Nagasaki University
58	Loewe	1987	Organ Dosimetry	DS86 Vol. 1:306–404 (Chapter 8)
59	Yamasaki	1987	Radioactive ^{32}P produced in sulfur in Hiroshima	DS86 Vol. 2:246–247 (Appendix 2 to Chapter 5)
60	Maruyama	1987	Composition of concrete from Joyama Primary School, Nagasaki	DS86 Vol. 2:248 (Appendix 3 to Chapter 5)

TABLE A-2 (*Continued*)

RefID	First Author	Year	Title	Journal
63	Okajima	1987	Quantitative measurement of the depth distribution of ^{152}Eu activity in rocks exposed to the Nagasaki atomic bomb	DS86 Vol. 2:256–260 (Appendix 6 to Chapter 5)
65	Shimizu	1987	Estimation of ^{32}P induced in sulfur in utility-pole insulators; at the time of the Hiroshima atomic bomb	DS86 Vol. 2:266–268 (Appendix 8 to Chapter 5)
67	Tajima	1987	Estimation of exposure dose	DS86 Vol. 2:269–271 (Appendix 9 to Chapter 5)
68	Hamada	1987	Measurements of ^{32}P in sulfur	DS86 Vol. 2:272–279 (Appendix 10 to Chapter 5)
71	Hasai	1987	^{152}Eu depth profile of stone bridge pillar exposed to the Hiroshima atomic bomb, Data acquisition of ^{152}Eu activities for the analysis of fast neutrons	DS86 Vol. 2:295–309 (Appendix 13 to Chapter 5)
73	Kimura	1986		Report to Monbusho pp 13–27
74	Miyajima	1986		Report to Monbusho pp 55–61
75	Nakanishi	1986		Report to Monbusho pp 62–72
76	Nakanishi	1986	Residual neutrons in Hiroshima	Draft report
77	Hoshi	1986	Motoyasu Bridge pillar	Draft (perhaps green book paper)
78	Nakanishi	1987		Isotope Center News Number 7 pp 2
79	Kerr	1987	Letter to Shigematsu August 14, 1987	Unpublished
80	Kato	1987	Europium isolation. . .	Anal. Sci. 3:493–497
81	Tajima	1988	Letter to Bill Ellett June 10, 1988	Unpublished
82	Brenner	1988	Neutron doses at Hiroshima	Columbia University Rad res annual report pp 61–64
83	Anon	1989	Notes from Irvine meeting	Unpublished
84	Kimura	1989	Steel bridge	Hawaii meeting
85	Maruyama	1989	Notes on Yokogawa Bridge	Hawaii meeting manuscript

(*continued*)

TABLE A-2 References in the RERF Dosimetry Measurements Database
(*Continued*)

RefID	First Author	Year	Title	Journal
86	Nakanishi	1989	^{152}Eu measurements	Hawaii meeting notes
87	Ruehm	1990	The neutron spectrum of the Hiroshima A-bomb and DS86	Nuc. Inst. Meth. Phys. Res. pp 557–562
88	Straume	1990	Use of accelerator mass spectrometry in the dosimetry of Hiroshima neutrons	Nuc. Inst. Meth. Phys. Res. pp 552–556
89	Shigematsu	1991	Japanese measurements	Letter to Bill Ellett 9-9-91
90	Straume	1992	Handout on ^{36}CL	Irvine meeting
91	Hoshi	1991	Studies of radioactivity produced by the Hiroshima atomic bomb: 1. Neutron-induced radioactivity measurements for dose evaluation	J. Radiat Res. Suppl. 20–31
94	Hoshi	1996		Proceedings of Nagasaki Symposium 50th anniversary pp 175
96	Okajima	1997	Nagasaki Eu measurements	1996 Report to Monbusho
105	Shizuma	1998	Residual ^{152}Eu and ^{60}Co activity induced by atomic bomb neutrons in Nagasaki	manuscript
106	Shizuma	1998	Residual ^{60}Co activity in steel samples exposed to the Hiroshima atomic bomb neutrons	Health Phys. 75:278–284
107	Nagatomo	1995	Thermoluminescence dosimetry of the Hiroshima atomic-bomb gamma rays between 1.59 km and 1.63 km from the hypocenter	Health Phys. 69:556–559
108	Nagatomo	1992	Comparison of the measured gamma ray dose and the DS86 estimate at 2.05 km ground distance in Hiroshima	J. Radiat. Res. 33:211–217
109	Ichikawa	1987	Thermoluminescence dosimetry of gamma rays from the Hiroshima atomic bomb at distances of 1.27 to 1.46 kilometers from the hypocenter	Health Phys. 52:443–451
110	Uehara	1988	Monte Carlo simulations of doses to tiles irradiated by ^{60}Co and ^{252}Cf simulating atomic bomb gamma-ray fluences	Health Phys. 54:249–256

TABLE A-2 (*Continued*)

RefID	First Author	Year	Title	Journal
111	Hoshi	1989	Thermoluminescence dosimetry of gamma rays from the Hiroshima atomic bomb at distances of 1.91–2.05 km from the hypocenter	Health Phys. 57:1003–1008
112	Haskell	1987	Thermoluminescence measurement of gamma rays—report on University of Utah analyses	DS86, Vol. 2:153–169
113	Ichikawa	1966	Thermoluminescence dosimetry of gamma rays from the atomic bombs in Hiroshima and Nagasaki	Health Phys. 12:395–405
114	Ichikawa	1987	Thermoluminescence measurement of gamma rays by the quartz inclusion method	DS86 Vol. 2:137–144
115	Nagatomo	1988	Thermoluminescence dosimetry of gamma rays from the atomic bomb at Hiroshima using the predose technique	Radiat. Res. 113:227–234
116	Nagatomo	1991	Thermoluminescence dosimetry of gamma rays using ceramic samples from Hiroshima and Nagasaki: A comparison with DS86 estimates	J. Radiat. Res. 32 (Suppl.):48–57
118	Maruyama	1987	Thermoluminescence measurements of gamma rays (Chapter 4)	DS86 Vol. 1:143–184
119	Roesch	1987	US-Japan joint reassessment of atomic bomb radiation dosimetry in Hiroshima and Nagasaki: final report (Vol. 1)	DS86 Vol. 1
120	Roesch	1987	US-Japan joint reassessment of atomic bomb radiation dosimetry in Hiroshima and Nagasaki: final report (Vol. 2)	DS86 Vol. 2
121	Maruyama	1987	Reassessment of gamma-ray doses using thermoluminescence measurements	DS86 Vol. 2:113–124 (Appendix 1 to Chapter 4)
122	Ichikawa	1987	Thermoluminescence measurement of gamma rays	DS86 Vol. 2:125–136 (Appendix 2 to Chapter 4)
123	Ichikawa	1987	Thermoluminescence measurement of gamma rays: quartz inclusion method	DS86 Vol. 2:137–144 (Appendix 3 to Chapter 4)

(*continued*)

TABLE A-2 References in the RERF Dosimetry Measurements Database
(*Continued*)

RefID	First Author	Year	Title	Journal
124	Nagatomo	1987	Thermoluminescence measurement of gamma rays by the pre-dose method	DS86 Vol. 2:145–148 (Appendix 4 to Chapter 4)
125	Hoshi	1987	Thermoluminescence measurement of gamma rays at about 2000 m from the hypocenter	DS86 Vol. 2:149–152 (Appendix 5 to Chapter 4)
126	Haskell	1987	Thermoluminescence dosimetry of atomic bomb gamma rays: University of Utah analyses	DS86 Vol. 2:153–169 (Appendix 6 to Chapter 4)
127	Eagleson	1987	Report from the Armed Forces Radiobiology Research Institute concerning LINAC and ^{60}Co irradiations	DS86 Vol. 2:169–170 (Appendix 6a to Chapter 4)
128	Hoffman	1987	Report on calibration and irradiation of samples with the UDM ^{137}Cs beam irradiator at the University of Utah	DS86 Vol. 2:170–171 (Appendix 6b to Chapter 4)
129	Bailiff	1987	Thermoluminescence analyses of Hiroshima ceramic tile and Nagasaki brick using the pre-dose and inclusion techniques	DS86 Vol. 2:172–183 (Appendix 7 to Chapter 4)
130	Huxtable	1987	Conventional thermoluminescence characteristics of a Hiroshima tile and a Nagasaki brick	DS86 Vol. 2:184–189 (Appendix 8 to Chapter 4)
131	Stoneham	1987	Thermoluminescence results on slices from a Hiroshima tile UHFSFT03	DS86 Vol. 2:190–197 (Appendix 9 to Chapter 4)
132	Haskell	1987	Interlaboratory calibration using NBS-irradiated Mg_2SiO_4:Tb	DS86 Vol. 2:198–203 (Appendix 10 to Chapter 4)
134	Thompson	1983	US-Japan joint workshop for reassessment of atomic bomb radiation dosimetry in Hiroshima and Nagasaki	First 1983 RERF Workshop
135	(RERF)	1983	Second US-Japan joint workshop for reassessment of atomic bomb radiation dosimetry in Hiroshima and Nagasaki	Second 1983 RERF Workshop
136	Ichikawa	1983	Thermoluminescent dating and its application to gamma ray dosimetry	First 1983 RERF Workshop: 104–114

TABLE A-2 (*Continued*)

RefID	First Author	Year	Title	Journal
137	Hoshi	1983	Thermoluminescent dating and its application to gamma ray dosimetry	First 1983 RERF Workshop: 115–121
138	Maruyama	1983	Reassessment of gamma ray dose estimates from thermoluminescent yields in Hiroshima and Nagasaki	First 1983 RERF Workshop: 122–137
139	Ichikawa	1983	Measurement of gamma ray dose from the atomic bomb by the quartz inclusion technique	Second 1983 RERF Workshop: 30–31
140	Haskell	1983	The use of thermoluminescence analysis for atomic bomb dosimetry: estimating and minimizing total error	Second 1983 RERF Workshop: 32–44
141	Maruyama	1983	Preliminary measurements of thermoluminescent yield with samples irradiated indoors	Second 1983 RERF Workshop: 45–47
142	Lowder	1983	Rapporteur's report	Second 1983 RERF Workshop: 48–51
143	Bond	1982	Reevaluations of Dosimetric Factors: Hiroshima and Nagasaki. Proceedings of a Symposium held at Germantown, Maryland, September 15–16, 1981	Germantown Conference Proceedings
144	(NCRP)	1988	Proceedings of the Twenty-third Annual Meeting of the National Council on Radiation Protection and Measurements: New Dosimetry at Hiroshima and Nagasaki and Its Implications for Risk Estimates	NCRP Proceedings No. 9
145	Hamada	1988	Early work carried out by Japanese scientists	NCRP Proceedings No. 9:5–13
146	Roesch	1988	Historical perspectives	NCRP Proceedings No. 9:14–22
147	Christy	1988	Overview of the new dosimetry: the physical basis	NCRP Proceedings No. 9:23–28
148	Haskell	1988	The use of thermoluminescence	NCRP Proceedings No. 9:32–48
149	Kosako	1988	Neutron activation studies related to the reassessment of Hiroshima and Nagasaki atomic-bomb dosimetry	NCRP Proceedings No. 9:49–63

(*continued*)

TABLE A-2 References in the RERF Dosimetry Measurements Database (*Continued*)

RefID	First Author	Year	Title	Journal
150	Kaul	1988	An assessment of dosimetry system 1986 (DS86) components	NCRP Proceedings No. 9:64–88
151	Kerr	1988	Sulfur activation in Hiroshima	NCRP Proceedings No. 9:99–106
152	Loewe	1988	Perspectives on radiation dose estimates for A-bomb survivors	NCRP Proceedings No. 9:107–116
153	Whalen	1988	Source spectrum and output spectrum calculations	NCRP Proceedings No. 9:117–120
154	Woolson	1988	The dosimetry system 1986 (DS86)	NCRP Proceedings No. 9:123–135
155	Preston	1988	The use of DS86 for the computation of dose estimates for Japanese A-bomb survivors	NCRP Proceedings No. 9:136–149
156	Higashimura	1963		Science 139:1284
157	Shizuma	1997	^{152}Eu depth profiles in granite and concrete cores exposed to the Hiroshima atomic bomb (1997)	Health Phys. 72:848–855
158	Shizuma	1997	Identification of ^{63}Ni and ^{60}Co produced in a steel sample by thermal neutrons from the Hiroshima atomic bomb	Nuclear Inst. Meth. A 384:375–379 (1997)
159	Fujita	1996	Exposed materials possessed by RERF which can be made available for TLD and neutron measurements	Report to Dosimetry Committees at Irvine, CA, Meeting
160	Nakanishi	1996	Recent improvements in radiochemical procedure for determination of ^{152}Eu at extremely low level	Report to Dosimetry Committees at Irvine, CA, Meeting
161	Maruyama	1996	Summary of thermoluminescence dosimetry measurements in Hiroshima and Nagasaki	Report to Dosimetry Committees at Irvine, CA, Meeting
162	Kosako	1996	Compilation of experimental dosimetry data for atomic bomb dose reassessment	Report to Dosimetry Committees at Irvine, CA, Meeting
163	Iimoto	1996	Measurement of ^{152}Eu induced by atomic bomb neutrons in Nagasaki	Report to Dosimetry Committees at Irvine, CA, Meeting
164	Iimoto	1999	Improved accuracy in the measurement of ^{152}Eu induced by atomic bomb neutrons in Nagasaki	Rad. Prot. Dos. 81 (2): 141–146 (1999)

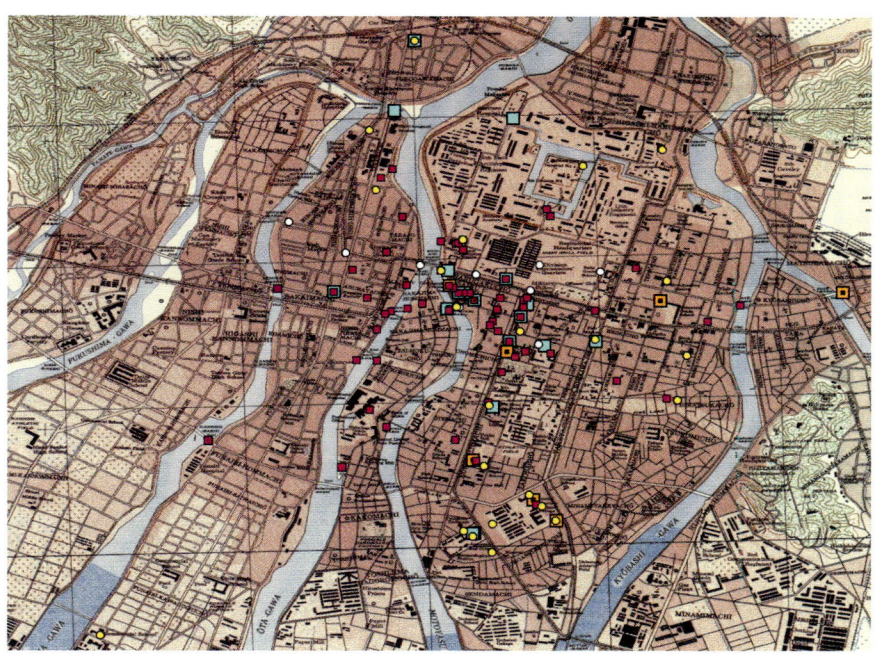

PLATE 1 Locations of thermoluminescent-dosimetry measurements in Hiroshima.

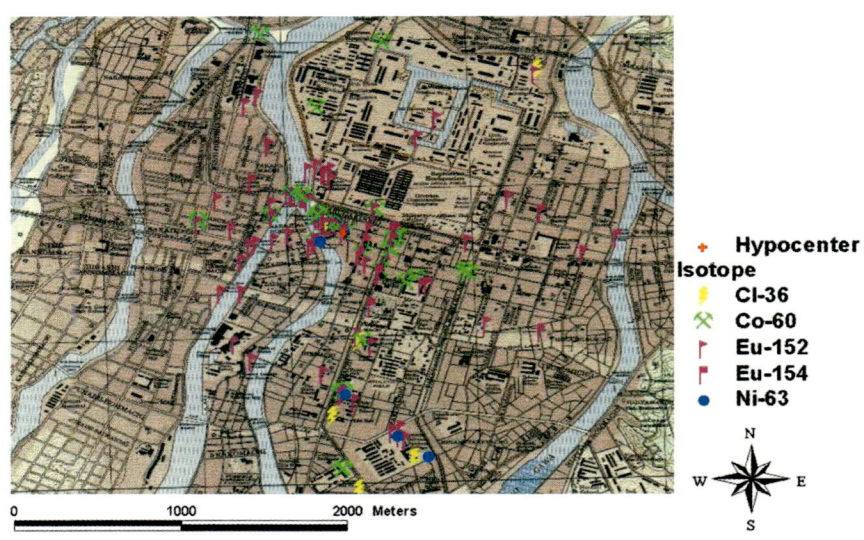

Hypocenter
Isotope
Cl-36
Co-60
Eu-152
Eu-154
Ni-63

0   1000   2000 Meters

PLATE 2 Locations of neutron-activation measurements in Hiroshima.

TABLE A-2 *Continued*

RefID	First Author	Year	Title	Journal
165	Maruyama	1988	Reassessment of gamma doses from the atomic bombs in Hiroshima and Nagasaki	Rad. Res. 113:1–14 (1988)
166	Hoshi	1992	Benchmark test of transport calculations of gold and nickel activation with implications for neutron kerma at Hiroshima	Health Phys. 63 (5): 532–542 (1992)
167	Kato	1988	Measurements of neutron fluence from the Hiroshima atomic bomb	J. Radiat. Res., 261–266 (1988)
168	Blamart	1992	Oxygen stable isotope measurements on a gravestone exposed to the Hiroshima A-bomb explosion and the "Dosimetry System 1986"	Chemical Geology (Isotope Geoscience Section) 101:93–96 (1992)
169	Ruehm	1992	^{36}Cl and ^{41}Ca depth profiles in a Hiroshima granite stone and the Dosimetry System 1986	Z. Phys. A—Hadrons and Nuclei 341:235–238 (1992)
170	Ruehm	1995	Neutron spectrum and yield of the Hiroshima A-bomb deduced from radionuclide measurements at one location	Int. J. Radiat. Biol.68 (1): 97–103 (1995)
171	Nakanishi	1998	Specific radioactivity of europium-152 in roof tiles exposed to atomic bomb radiation in Nagasaki	J. Radiat. Res., 39: 243–250 (1998)
172	Endo	1999	DS86 neutron dose: Monte Carlo analysis for depth profile of ^{152}Eu activity in a large stone sample	J. Radiat. Res., 40: 169–181 (1999)
173	Ito	1999	A method to detect low-level ^{63}Ni activity for estimating fast neutron fluence from the Hiroshima atomic bomb	Health Phys. 76(6): 635–638 (1999)
174	Kimura	1993	Determination of specific activity of ^{60}Co in steel samples exposed to the atomic bomb in Hiroshima	Radioisotopes 41:17–20 (1993)
175	Maruyama	1999	Determinations of background in the pre-dose TL technique	draft manuscript, personal communication from Dr. Maruyama 12-13-99
176	Straume	2000	Neutron measurement update	Notes from Dosimetry Workshop, Hiroshima, 13–14 March, 2000

(continued)

TABLE A-2 References in the RERF Dosimetry Measurements Database (*Continued*)

RefID	First Author	Year	Title	Journal
177	Shizuma	2000	Residual radioactivity measurement in Hiroshima and Nagasaki for the evaluation of DS86 neutron fluence	Poster at IRPA 10, Hiroshima, May, 2000
178	Shizuma	1999	Contribution of background neutron activation in the residual activity measurement and present status of ^{152}Eu measurements for Nagasaki samples	Notes from binational meeting on RERF dosimetry, Irvine, CA, January 1999
179	Goldhagen	1996	Neutron spectrum measurements at distances up to 2 km from a uranium fission source for comparison with transport calculations	Proceedings of the American Nuclear Society Topical Meeting, April 21–25, 1996
180	Maruyama	2000	Summary of thermoluminescence measurements in Hiroshima and Nagasaki	U.S.-Japan Joint Dosimetry Workshop, March 13–14, 2000, Hiroshima, Japan

QUESTIONNAIRE

This questionnaire prepared by W. Lowder and T. Maruyama of the U.S. and Japanese dosimetry committees, is designed to provide a basis for the collection of important information with regard to each sample of environmental material analyzed for neutron activation or gamma-induced thermoluminescence at Hiroshima and Nagasaki. Its purpose is to indicate the key questions that will be addressed during the visits of Dr. Maruyama and Mr. Lowder to the various laboratories in the U.S. and Japan where relevant measurements and calculations have been made. The individual investigators can make use of this questionnaire to prepare for those visits and have the needed information readily available at the time.

The information gathered will be used to conduct an uncertainty analysis designed to identify and quantify those factors that contribute to the overall uncertainties of both measurements and calculations. The term "uncertainty" refers to both precision and accuracy, involving questions of reproducibility and bias. It can be expressed in terms of confidence limits, probable errors, standard deviations, etc.

The questionnaire is divided into four sections. All investigators should review section A, which uniquely identifies the subject samples. Since different investigators are often involved in the various aspects of the collection, processing, and measurement of the samples and the conduct of the associated calculations, only those later sections that pertain to the work done at your laboratory need be considered. However, it is essential that each sample as measured can be unambiguously related to a particular field sample as collected and to a particular fluence calculation at the location of collection. So particular attention should be paid to those questions relating to sample and subsample ID's, origin, transfer between laboratories, and current status, as well as relevant calculations. Note that some questions are repeated in different sections, so that each section is self-contained.

Section A: Basic Information

(1)   Provide name of responder and institution.

(2)   Provide ID of sample(s), type of material, and a brief description.

(3)   Indicate field sample collection location (city, structure, distance and direction from hypocenter).

(4)   Indicate type of measurement, e.g., "thermal neutron activation, ^{152}Eu" or, "T quartz."

Section B: Field Sample Collection and Treatment

(1)   Provide field sample ID as assigned by the collector.

(2)   Provide date of collection and name of responsible investigator.

(3)   Provide a brief description of the sample as collected, including type of material, size, and weight.

(4)   Describe the site of collection, including the structure containing the samples, local terrain (water and ground), and overall structural shielding geometry associated with nearby structures (to define the immediate environment surrounding the sample that affects the calculations).

(5)   Give the age of the structure containing the sample and of the sample, if different.

(6)   Give the height above ground of the sample collection point.

(7)   Give the sample orientation relative to the line-of-sight to the burst.

(8) Provide the distance and direction from the hypocenter as determined by the collector, and indicate the method used for this determination.

(9) Provide an estimate of the uncertainty in the distance determination and indicate the method used.

(10) Describe any treatment of the field sample, including the division into subsamples.

(11) Provide the ID's of any subsample, as assigned by the collection library.

(12) Indicate the disposition of sample and subsamples, including when, where, and to whom they were sent.

(13) Describe the current status of any sample or subsample retained at the collection laboratory.

Section C: Measurement Sample Preparation and Measurement

(1) Indicate sample or subsample ID's as received (collector's and/or investigator's).

(2) Give date received and from whom.

(3) Describe the sample or subsample(s), including location of collection and field sample ID.

(4) Indicate the origin of the sample or subsample(s), including both location of collection and field sample ID.

(5) Describe sample treatment procedures to prepare measurement sample(s), e.g., further division, chemistry.

(6) Provide any information on sample composition, how the composition was determined, and the source of such information.

(7) Indicate the position of measurement sample in collected field sample, if known.

(8) Provide ID's of each measurement sample and date of measurement.

(9) Describe briefly the method of measurement, including calibration procedures.

(10) Indicate whether calibration factors have been checked by means of inter-comparisons with other laboratories or other methods of quality assurance.

(11) Give the direct result(s) of the measurement(s) and its uncertainty, indicating exactly what was measured.

(12) Indicate how the measurement uncertainty was determined.

(13) Describe how the measurement background was determined (e.g., contributions from contamination, natural radiation).

(14) Describe the conversion of the measured quantity to the desired quantity. (give numerical values and units for all quantities in this conversion, including background).

(15) Give the final result(s) (that is, the eventual "M" in the C/M determination), with uncertainty and units.

(16) Indicate how this uncertainty was determined.

(17) Indicate the calculation result(s) used to determine the C/M ratio(s) (give source and reference for such results).

(18) Give the final result(s) for the C/M ratio, with estimated uncertainty.

(19) Describe how the uncertainty in the C/M ratio(s) was determined.

(20) List all published papers, laboratory reports, and reports to the dosimetry committees that include and discuss these particular measurements.

Section D: Calculations

(1) Describe briefly the field sample collection site (city, structure, distance and direction from the hypocenter), and indicate if an independent estimate was made of distance from the hypocenter (with value and uncertainty).

(2) Provide field and/or measurement sample ID's.

(3) If different from DS86, describe briefly how free-air neutron or gamma fluences at the field sample collection site were determined.

(4) Describe how the sample response was calculated. Include information on how the collection site was modeled (e.g., sample location and properties, structures, local media) and how the radiation transport from free air to the sample location was determined.

(5) Give the calculated values, with uncertainties, of free-air fluences, fluence at the sample location, and sample response.

(6) Indicate how the uncertainties were determined.

List known published papers, laboratory reports, and reports to the dosimetry committees that include and discuss these calculated values and associated C/M ratios.

Appendix B

An Uncertainty Analysis of Neutron Activation Measurements in Hiroshima and Nagasaki

Neutron activation was measured with two fundamentally different types of detection with different types of errors: counting atoms with radiometric methods (beta or gamma counting) and counting atoms with accelerator mass spectroscopy (AMS). The following discussion focuses first on the general approach to error analysis that is used here, and then on each classification of measurement.

The general approach used here is to begin by exhaustively identifying all the quantities that are measured and the formulas by which investigators use the quantities to calculate results in the terms commonly reported in the literature. There must initially be a thorough consideration of the uncertainty in all the measured *or assumed* values of quantities that are used by investigators to arrive at the final reported result, although some of the quantities may prove to be known with such accuracy and precision that their uncertainty can be ignored in a quantitative treatment. The latter quantities can then be treated as constants. For example, we assume the weighing of samples and standards to be so precise and well calibrated that the associated error is not included in any of the estimates calculated here. But, a volumetric error is typically associated with pipetting microliter quantities of liquids, such as might be used to prepare calibration standards, and this would be a potential contributor to experimental error of some significance if not verified and corrected by weighing.

The objective is to estimate the total uncertainty of each reported result in relation to some presumed true value of interest. *For the purposes of this appendix, the true value of interest is the amount of the neutron activation product nuclide per unit mass of the associated stable target element that existed in the sample being measured at the time of the bombing (ATB) in 1945.* Depending on the measurement method, this value may be stated in units involving ratios of numbers of atoms or in units of radioactivity per unit mass of the target element.

Sample-specific activity ATB (August 6 or 9, 1945) is related to the dosimetric quantities of interest for survivor dose in a complicated way. The sample-specific activity ATB is a function of the bomb-related thermal-neutron and epithermal-neutron fluences that existed in the sample ATB. Those fluences are in turn related to the free-field neutron fluences at the same location in a way that depends to some extent on local moderation and absorption of the free-field neutron fluences. Local moderation occurs in the terrain and structures close to the sample and in the sample itself. The free-field neutron fluences are of paramount interest because they are the quantities used to calculate doses to survivors and because they constitute a uniform basis for the comparison of measured values. This appendix does not analyze the uncertainty in the free-field neutron fluences calculated by DS86; they are discussed in Chapter 6 of this report. The relationship between free-field and in-sample neutron fluences is discussed below in connection with the plotting and fitting of measurements.

For analytical purposes, every measured or assumed value is treated here as a random variable. The difference between a true value and an individual measured value has both a systematic and a random component, corresponding to a mean difference and a standard deviation. The former is commonly characterized as a bias and the latter as a random error. Every experiment is based on methods that are intended to minimize systematic and random errors. Although it is difficult for a retrospective analysis as reported here to obtain sufficient information to identify and provide a useful quantitative estimate of a nonzero bias, every reasonable effort is made to do so. More often, the main quantitative result of this type of analysis is to verify the magnitude of an investigator's estimated random error or provide a more realistic estimate of the true random error. This type of analysis can also help to identify possible sources of error that cannot be quantified with the information at hand but can be addressed in recommendations for future work.

Another way to look at the uncertainty issue is to focus on the limiting case of measurements that approach the limits of detectability by making generic calculations of limits of detection with accepted formulas based on statistics. For radiation-counting methods, such calculations must be based on assumed nominal values for sample and background counting time, counting efficiency, and the amount of the stable target element present in the sample, which are treated as constants for the purpose of the calculation. Those calculations are useful for illustration and for defining limits below which reported measured results should be treated with particular caution. They also help to define the nature of some important relationships for the lowest-level measurements, relating the comparative sizes of counting-system background, sample background, and calculated or measured sample content from the bomb fluence. These issues are addressed in detail below.

PROPAGATION OF ERROR

To estimate the uncertainty in the final reported result, sample-specific activity ATB, a propagation-of-error calculation can be based on the equations by which the measured and assumed values of quantities are used by investigators to calculate reported results, incorporating the uncertainty estimates of the individual values involved. The desired result is an estimate of the total uncertainty in the final reported result that is due to all sources combined. For the purpose of this report, standard formulas based on the first-order terms of Taylor-series expansions are used. It should be noted that these formulas tend to underestimate somewhat the variance in the reported values as a function of the available estimates of component uncertainties. The underestimation might not be negligible. As one indication of the potential inadequacy of first-order approximations, it can be noted that a large majority of the major-component uncertainties given by measurers have fractional standard deviations exceeding 10% in one or both components typically reported. To evaluate and improve the estimates of uncertainty in the specific activity ATB, numerical simulations were performed.[1] For any individual result, there is generally no reason to believe that errors in the different components are correlated. The resulting propagation-of-error calculations are therefore relatively simple; the formulas used in spreadsheet calculations based on the first-order approximation are given in the sections below. In general, it is noted that for any set of uncorrelated values and estimated standard deviations, say,

$$X \pm a, Y \pm b, Z \pm c$$

and any constants m_i, the sum or difference formula gives the result

$$SD(m_1 X \pm m_2 Y \pm m_3 Z) \cong \sqrt{m_1^2 a^2 + m_2^2 b^2 + m_3^2 c^2} \qquad (1)$$

---

[1] When simplified numerical simulations (multivariate normal with zero covariance) are performed to evaluate the combined error of sample-specific activity as calculated from a radiation counting, it is notable that if any term in the divisor of the equation as it is normally formulated (counting efficiency, result of the stable element assay) begins to exceed about 12–15% coefficient of variation, the error distribution of the specific activity becomes badly skewed upward. That is because such a situation involves a nontrivial probability of a small value close to zero in the divisor of the formula for specific activity, with such small values causing arbitrarily large calculated specific activities. Several of the [152]Eu measurements of Shizuma and others (1993) have rather large estimated error in the assay of stable europium as reported by the authors. Of these, two cases may deserve review of the determined stable europium content because the calculated specific activity is considerably larger than other nearby measurements: Sorazaya Shrine, 873-m slant range, 15% estimated SD in stable europium, and Enryu Shrine, 1081-m slant range, 14% estimated SD in stable europium. In addition, it should be noted with regard to all the radiometric neutron-activation measurements in the literature that if there are cases in which the author's estimates of the error in stable europium and cobalt assays are substantially understated, those measurements might be similarly affected.

and the product or quotient formula gives

$$\frac{SD\left(\dfrac{mXY}{Z}\right)}{\dfrac{mXY}{Z}} \cong \sqrt{\left(\frac{a}{X}\right)^2 + \left(\frac{b}{Y}\right)^2 + \left(\frac{c}{Z}\right)^2} \tag{2}$$

As a matter of nomenclature, the *coefficient of variation* σ/μ (standard deviation divided by mean) for a variable of interest may be given herein as a fraction and called a *fractional standard deviation,* or it may be given as a percentage and simply called a *% error.* Thus, Equation 1 says that for such formulas as sums, differences, and weighted sums the standard deviations themselves sum in quadrature. Equation 2 similarly says that for products and quotients the *fractional* standard deviations sum in quadrature.

CORRELATED ERRORS

In addition to the possible effect of correlation of the errors in the component quantities on a single result, another type of correlation must be considered in any application based on more than one measured result: correlation of errors among reported results. Various subsets of measured values, classified at various levels (for example, within the same sample, within a given sampling location, within a given investigator's laboratory, and within a given range of calendar time), can share the same measured or assumed value for some part of the calculated result, such as a calibration factor, and can therefore be correlated with respect to that factor. Such cases will be discussed in detail.

One might ask whether the measurements at a particular site can tend to share a common bias relative to the true value, that is *not* due to sharing a common value for something, such as a calibration factor. To the extent that such a covariance might exist, it would most likely be due to an unmeasured covariable that affects the true value for the sample, rather than to an error inherent in the measurement process. Nothing about the site should affect the process, and the properties of the sample should have minimal effect on the measurement process. For example, the sample-specific properties that would affect the counting efficiency, such as the effect of the elemental composition of the sample on its self-absorption of the emitted radiation being measured, are likely to have negligible influence.

However, there might be sample-specific variables that appreciably affect the neutron activation level of the sample for a given incident bomb fluence (such as boron content or water content) that have not been measured or have not been properly incorporated in the calculated value for the sample. Such site-correlated errors are not included in the uncertainty analysis reported here, because they are errors in the calculated value and not the measured one as defined here. However, they do need to be considered in fitting curves to the measurements. They are discussed further in the next section.

COMBINED UNCERTAINTY ESTIMATES: ^{32}P MEASUREMENTS OF FAST NEUTRON FLUENCE

The ^{32}P measurements made by Japanese physicists in 1945 (Yamasaki and others 1987) have often been cited in relation to general source term parameters such as yield and height of burst. Because these measurements are at fairly close distances and are used in this manner, it is of some interest to characterize their uncertainty as an aggregate, as well as individually. Unfortunately, characterizing the uncertainty of these measurements in the aggregate, or using them in fitting a model, is difficult to do correctly, because their errors are highly correlated. Even the calculation of a weighted mean for these measurements, for the reasons discussed below, would require a very careful and somewhat complicated approach to propagation of error.

The ^{32}P measurements were very fortuitous, or very well planned, in several respects, including the facts that the measurements were made on essentially pure elemental sulfur of reliably high purity, and the measurements were originally calibrated with a natural radioactive source of good chemical purity, uranium oxide, whose emission rate of beta particles can therefore be accurately predicted. These factors, along with the preservation of the Lauritsen electroscopes used to perform the measurements, allowed a series of careful retrospective studies of factors related to the measurements' accuracy by Hamada (1983a,b, 1987) and Shimizu and Saigusa (1987). Hamada (1987) estimates a 2% random error in uniformity of sample preparation that relates to a counting efficiency factor, presumably related primarily to the evenness of spreading the powdered sample on the glass plate for counting and the resulting differences in self-absorption of betas. Based on the accuracy quoted by Hamada (1987) for his ^{32}P reference standard and the likely counting error variance of his calibration measurements, it would seem reasonable to estimate that the calibration error should not exceed 5% or so, where the greater error would likely be the counting error of his calibration measurements rather than the accuracy of his standard reference material.

These numbers can be combined with the counting error given by Hamada (1987) in his Table 2 to provide estimates of the errors in individual measurements considered in isolation, which are shown in Table B-1. (The chemical purity of each sample, as it relates to the accuracy of the estimated weight of pure sulfur present in each sample, appears to be high and to have a fractional standard deviation of only a fraction of a % based on Hamada's data (1987). It is ignored in this calculation.)

These estimates differ very little from those of Hamada, because the larger counting errors do in fact predominate. The fractional standard deviation (FSD's) of individual measurements are fairly large, up to 53%. Some measurements with even larger errors are not shown in the table, and some measurements for which the azimuth is unknown are also omitted.

TABLE B-1 Estimates of Errors in Individual Measurements Considered in Isolation

Sample	Counting Interval, min.	Electroscope Reading, div s^{-1}	Activity (Net) dis s^{-1} Sg^{-1}	Total Error, SD dis s^{-1} Sg^{-1}	Total Error % FSD[a]
background	200	0.00124	—	—	
A4	40	0.00200	2200	434	20
B5	38	0.00224	2940	441	15
C6	45	0.00154	880	399	45
D12	70	0.00162	1500	442	29
D12	70	0.00178	1140	247	22
E13	95	0.00205	2430	342	14
F14	74	0.00166	1260	335	27
G15	90	0.00167	1370	337	25
G15	90	0.00175	1620	336	21
H7	85	0.00144	630	336	53

[a] FSD = fractional standard deviation.

An important caveat is that the errors in these measurements are highly correlated. This is because all measurements appear to share a common determination of background, which is an additive error, and a common determination of counting efficiency, which is a multiplicative error. Background in particular was a rather proportionally large error in this experiment. Any effort to use these error estimates in fitting to calculated values should involve a very careful propagation of error formulation that begins by extracting the common error in background. This would require a careful reestimation based on reworking Hamada's (1987, actually, Roesch's and Jablon's) equations (1) through (8) for calculating error in count rate based on error in time to reach a common electrical charge on the electroscope.

INVESTIGATORS' ESTIMATES OF UNCERTAINTY BASED ON COUNTING STATISTICS

In radiation-counting applications, investigators have almost universally calculated their estimated errors in the radioactivity content of samples on the basis of Poisson counting statistics. With rare exceptions, these appear to be the sole basis of plotted error bars and published estimates that are intended to suggest the precision of measured values. But the raw data of the measurements are generally not available to allow checking of the calculations. For example, in the case of radiometric methods, one would have to obtain at least the

- Calendar date(s) when samples were counted.
- Lengths of the counting intervals (in detector live time) and counts in the region of interest and background subtracted, for all counts of bomb-fluence samples.
- Blanks for background and calibration standards for counting efficiency.

Although it was the intent of the committee's questionnaire (see Appendix A) to obtain such information, it was not provided by most investigators.

The most basic aspect of the calculations can be checked in a limited way via a calculation of minimum detectable activity (MDA). If one has estimates of background count rate, counting efficiency, and counting interval for sample and background, an MDA can be calculated and compared to give the sample's estimated content. Such calculations have the limitation that they assume the counting efficiency and background count rate as fixed constants. The validity of such comparisons is also a function of the extent to which the assumed counting intervals are representative of those used. Calculations of MDA and related quantities are discussed further in later sections.

ESTIMATION OF COMPONENTS OF UNCERTAINTY NOT ESTIMATED BY INVESTIGATORS

Because the method of uncertainty analysis was limited to what could be done with incomplete information, it was necessary that the method be flexible and carefully adapted to each individual situation. In some cases, particular uncertainty components have been estimated on the basis of expert judgment and knowledge of typical standards and practices or by using a carefully considered application of values obtained by other investigators with similar methods. Where such judgments have been made, they are clearly so identified in the following sections, regardless of the magnitude of their effect (sensitivity) on the total uncertainty estimates.

STATISTICAL DISTRIBUTIONS OF ERRORS

A word about statistical distributions is in order. The types of detection methods considered here are counting methods, and their raw results are expected to obey the Poisson distribution. At present it appears that all measurements have sufficient numbers of counts for the Poisson to be well approximated by the Gaussian distribution; therefore, skewness is not a major concern. The counting statistics tend to dominate the uncertainty of the measurements, especially inasmuch as the assays were typically calibrated by comparison with other results of the same (radiation-counting) type applied to standard materials. In the case of radiation measurements, assay of stable-element concentration is another major factor in the reported result. Radiation counting of the activation product radionuclide is a third. Most assays of elemental concentration were also performed with radiation counting after controlled neutron irradiation. Moreover, reported results are usually based on averaging of several measurements, and the central limit theorem supports the distributions of component and overall errors' being approximately Gaussian.

Statistical distribution does not affect the second-moment properties that define the propagation of error equations, but it does affect the interpretation of a standard deviation in terms of cumulative probability. For example, the construction of confidence intervals depends on the type of the assumed underlying error distribution and on the estimated value of its standard deviation. Another important consideration in this particular analysis is that the type of error distribution affects decisions about the type of transformation that should be applied to the variable, if any, to fit curves to the data as a function of distance from the epicenter.

ERROR IN THE INDEPENDENT VARIABLE: DISTANCE FROM THE HYPOCENTER

There is appreciable error in the principal independent variable of interest in all analyses of the measurements: the distance from the burst point (epicenter) of the bomb to the location of the sample. That distance is used as a major input for calculating survivor dose. The standard approach in all dose-calculating systems has necessarily been to assume radial symmetry about the hypocenter (ground zero) of the bomb. Therefore, the free-field value calculated by the dosimetry system—the value for an idealized infinitesimal-volume element of air or tissue suspended 1 m above flat ground—depends only on the radial distance from the epicenter (slant distance) or hypocenter (ground distance).

Most investigators have published estimates of uncertainty in distance with their measurements; among neutron measurements, this information is lacking for only a small portion of the data, mainly values reported in the DS86 final report and earlier source documents. The values that are published by measurers are somewhat subjective and have some unusual attributes. For example, because of the geographical context, investigators have tended to think of the "plus-or-minus" values that they estimate as being something like a maximal credible range, rather than a standard deviation. The plus-or-minus uncertainty values tend to fall mostly in a range from about 3 m to about 30 m, but there are a few values as high as 90 m, for samples for which the measurer knew only that a sample came from a particular large building and could obtain no better information from the collector.

The magnitude of the effect of distance error on calculated neutron activation is a function of two factors: geometry and effective attenuation of the radiation fluence. The effective attenuation of the fluence by interactions in air and on the ground predominates: a relaxation length of 125 m, for example, corresponds to a change in fluence of roughly 25% over 30 m at any distance.

In contrast, at distances of interest in connection with survivor dose, the inverse square effect is small: it changes fluence from a point isotropic source by only about 6% over a slant distance of 30 m at 1 km from the epicenter, and 4% at 1.5 km. The effect is larger at shorter distances of interest for fitting the measurements, such as about 10% at 600 m. For a more extended source, such as the fireball, as might apply to delayed radiation to some extent, the geometrical dependence is closer to the inverse of distance than to the inverse square, and the effect is correspondingly smaller.

Efforts are in progress at RERF to provide improved estimates of map location, distance, and related uncertainty by using geographical information system (GIS) software in combination with the extensive source documents available. Distance uncertainty is discussed below in relation to fitting curves to the data.

GRAPHICAL DEPICTION AND FITTING
OF CURVES TO THE DATA

Scaling by the Inverse Square of Distance

For purposes of graphical depiction, measurement results stated ATB are plotted against slant distance from the epicenter on a semilogarithmic plot, which is standard in the literature. To facilitate visual comparisons, all values—those calculated by DS86 and those calculated from measurements—are multiplied by the square of slant distance in kilometers. That removes the inverse-square dependence on distance that is universal for radiation emitted isotropically from a point source. Hence, in the absence of attenuation, radiation fluences from such a source, so scaled, would have a perfectly constant fluence. And the fluences of radiation from such a source if subject to exponential attenuation as a function of distance, would fall on a straight line, whose slope is commonly characterized by its inverse in terms of the natural logarithm; the distance subject to attenuation by a factor of 1/e, is commonly called a *relaxation length.* Any systematic departure from a straight line indicates a departure from one or both of these assumptions: the source is not isotropic or is not of small extent compared with the distances involved, or the attenuation is not effectively an exponential function of distance.

Appropriate DS86-Calculated Values for Comparison
with Specific Measurements

To compare measured values with DS86-calculated values, it is necessary to determine appropriate DS86-calculated values. Samples were chosen by investigators to be near the surface of the sampled material in a location with a direct line of sight to the epicenter, with three exceptions:

• Samples that were deliberately taken at increasing depths in the material at a given location to measure activities related to depth.
• In the case of gamma thermoluminescent dosimetry measurements, samples like the pottery shards from the interiors of houses and buildings or underlying roofing tiles that were chosen with foreknowledge that the results would be questionable and that were prominently so identified as "shielded samples."
• Steel concrete reinforcement rods ("rebar") located at depths approximating 8 cm in concrete, that were measured in the 1960's with a specific plan in mind to evaluate factors related to the spectra of incident neutron fluences (Hashizume and others 1967).

The approach of exclusively measuring surface line-of-sight samples for evaluating neutron activation as related to distance was pursued in the belief that such samples would yield measured values as close as possible to the free-field or free-in-air value at the location in question. That approximation is subject to question. The extent to which a given sample reflects the free-field value at 1 m above flat ground, as a standardized reference value for systematic comparisons of measured values as a function of distance from the bomb, depends on more than its being a surface line-of-sight sample. It also depends on the size, shape, and composition of the structure in which the sample is situated and on the properties of the surrounding terrain.

The situation is further complicated in that DS86 was written to calculate not neutron activation, but rather neutron dose to tissue. Calculating neutron activation, even a free-field value, requires a mathematical convolution of the appropriate neutron-interaction cross-section values with the energy-dependent neutron fluences given by DS86. Such calculations should be done by an expert in any case.

The neutron measurements are in four categories with respect to the DS86-calculated values available for comparison:

• Measurements for which detailed calculations based on DS86 neutron fluences have used Monte Carlo or S_n simulations with a model of the structure containing the sample, such as the S_n calculations for the Yokogawa Bridge samples by Oak Ridge National Laboratory (Kerr and others 1990) or the calculations done for the Motoyasu Bridge pillar by Hasai and others (1987).

• Measurements for which relatively simple calculations have been done by Scientific Applications International Corp. (SAIC) to account for the shielding effect of materials overlying the sample.

• Measurements that have been reviewed by an expert at SAIC (Dr. Egbert) and classified as being well approximated by the free-field value on the basis of expert judgment regarding the nature of the sample location.

• Measurements, mostly published since 1997, for which no expert evaluation has been performed and the only value available for comparison is the free-field value.

M/C and C/M Plots vs. Plotting M and C Separately

For purposes of comparing measured and DS86-calculated values as a function of distance, some investigators have preferred simply to plot the ratio of measured to calculated (M/C) or the ratio of calculated to measured (C/M) values. However, that gives no information about the behavior of the two individual quantities as a function of distance. When plotting the quantities separately against distance, it is natural to plot the DS86-calculated free-field values rather than the sample-specific in situ calculated values because the former lie on a continuous curve. If the measured values are plotted without modification, their difference from the DS86-calculated free-field values reflects both

• The difference between a DS86 free-field value at the indicated distance and a completely calculated DS86-based value for the sample, where the latter includes the effect of local terrain and the shielding due to overlying materials, as would be calculated by a full application of computational methods to the DS86 free-field fluences.
• Any difference between the measured value and such a completely calculated value for the sample.

To make plots and fitted values that focus on the discrepancies between measured and DS86-calculated values, we made two key decisions:

• Measurements at subsurface depths in the sample material were omitted to minimize the shielding correction between free-field and sample-specific calculated values, and
• Plotted and fitted measured values were corrected for the ratio between the DS86 free-field calculated value and the most completely calculated sample-specific DS86 value available. That is, "measured free-field equivalent values" were calculated to remove the effect of shielding and local terrain as much as possible and obtain the free-field value that would presumably be associated with the in situ measured value.

It is emphasized that "measured free-field equivalent values" do not reflect any calculation that is new or different from what has been done before. They merely represent a way of plotting measurements in relation to a continuous curve for the DS86-calculated value as a function of distance for samples that have sample-specific calculated values that differ from the free-field calculated values.

Choice of Functions for Fitting to the Data

The DS86-calculated gamma dose for both cities falls very close to a straight line on a semi-log plot of values scaled by the square of distance: its attenuation is close to exponential, and its effective source size is fairly small relative to all distances on the ground. Some $1/r$ dependence at shorter distances is to be expected from the distributed nature of such sources as fission products in the fireball and neutron-capture gammas arising from interactions with nitrogen in the air near the explosion. Efforts to estimate coefficients separately for a $1/r$ dependence and a $1/r^2$ dependence in fits to the measured values were not successful, apparently because the rate of change in the fitted values due to a $1/r$ dependence is smaller that due to the exponential attenuation and because of the lack of precision in the measurements. The DS86-calculated neutron activation for both cities shows a small but appreciable upward curvature on such a plot: its relaxation length increases somewhat with distance. For Nagasaki, there is also a substantial departure from a straight line at distances less than about 600 m. For that reason, points representing less than 550-m slant distance in Nagasaki, both DS86-calculated and measured, have not been included in fitted curves. (This peculiarity is related to the source term, and is different for the 1993 suggested modifications to DS86.)

Inherent Assumptions of M/C and C/M Curves

Fitting curves through the data must be done with an understanding of the assumptions on which the fitted curves depend. In no case are such assumptions trivial or beyond question. For example, fitting a line through the ratios of DS86-calculated and measured values on a semilogarithmic plot against distance r is by definition a matter of fitting

$$\frac{M(r)}{C(r)} = \beta e^{\alpha r} \tag{3}$$

with α and β as fitted parameters. This is a natural result only if M and C are functions of distance that can be separated into a factor, some arbitrary function $f(r)$, that is identical for M and C and a factor that is exponential in r:

$$\frac{M(r)}{C(r)} = \frac{f(r)\beta_m e^{\alpha_m r}}{f(r)\beta_c e^{\alpha_c r}} = \beta e^{\alpha r} \tag{4}$$

with $\alpha = \alpha_m - \alpha_c$ and $\beta = \beta_m/\beta_c$. It is not a natural result in any more general context. If the fitting is done on logarithms with a simple linear regression and all measurements are given equal weight, there is no consideration of the large variability in the precision of the various measurements. In addition, a subtler but constraining decision is being made regarding the presumed nature of the error distribution: that it is approximately normal in the logarithmic transform, as would be the case for a *lognormal* distribution of error in the *untransformed* measured value.

Method of Fitting

To provide a physically meaningful comparison of measured values and DS86-calculated values, some curves must be fitted. Otherwise, there can be no discussion of parameters, such as relaxation length. To provide a basis for such comparisons, a decision was made to fit curves to the scaled DS86 free-field calculated values and scaled measured values corrected for cosmic background and for the ratio of the free-field value to the most completely calculated sample-specific in situ value available. We decided to work in the original space rather than using the logarithms or another transform of the values and to obtain fitted curves by using a nonlinear regression and weighted least squares. We fitted curves that were based on a simple nonlinear function that gives a good fit to the DS86-calculated values and is consistent with the physics of radiation transport, and we made an important decision to weight each observation by the inverse of its estimated variance as derived from the uncertainty analysis.

Two concerns were recognized in considering how to assign weights to the data to determine the amount of influence that each measurement would have on the fitted result. First, some weighting for the relative uncertainty of each type of measurement is desirable: measurements with greater uncertainty should have less influence. The other concern is related to the fact that the values of interest span a factor of 10^3–10^4 in some cases: values measured at the longest distances are sometimes 10^{-3} or 10^{-4} of values measured near the hypocenter. A residual of any given absolute size is 10^3 or 10^4 times as large in proportion to the average measurements at the longest distances as at the shortest distances. With no weighting, the residuals for the least-squares fit would tend to be that much larger as well in relation to the measured values. If all measurements had been made to an equal relative precision (equal estimated coefficients of variation), this would seem inappropriate.

A simple and natural solution is to use the raw uncertainties reported by the investigators as weights in an inverse-square formulation. This addresses both issues in an intuitively appealing way. That is, if the author reported a value as "$x \pm y$," where y is stated in the same units as x, then the weight used for that measurement in the regression would be simply

$$weight = \frac{1}{y^2} \tag{5}$$

Two important relationships are clarified by restating Equation 2 as a product of two implicit factors:

$$weight = \frac{1}{y^2} = \frac{1}{x^2} \cdot \frac{x^2}{y^2} \tag{6}$$

Weighting by $1/y^2$ thus amounts to using a weight equal to the product of the inverse of the square of the measurement itself and the inverse of the square of an estimate of the coefficient of variation, σ/μ, of the measurement. Thus, x estimates μ, and y estimates σ.

Inasmuch as the first factor is inversely proportional to the square of the measurement, smaller measurements have higher values of this factor to make up for the lower size that their residuals would tend to have in a proper fit to the correct functional form with the same relative measurement precision. That is, among measurements of the same estimated relative precision, stated as estimated coefficient of variation (same value of y/x), the second factor is constant, and the weights are inversely proportional to the squares of the measured values. Under these conditions, any discrepancy between fitted and measured values that is some given proportion of the measured value produces a term of the same size in the weighted sum of squares:

$$\frac{x_1}{y_1} = \frac{x_2}{y_2} \cap \frac{x_1 - A(r_{x_1})}{x_1} = \frac{x_2 - A(r_{x_2})}{x_2} \Rightarrow \left(\frac{x_1 - A(r_{x_1})}{x_1}\right)^2 \left(\frac{x_1}{y_1}\right)^2 = \left(\frac{x_2 - A(r_{x_2})}{x_2}\right)^2 \left(\frac{x_2}{y_2}\right)^2 \quad (7)$$

where $A(r_x)$ is the fitted curve according to Equation 1 evaluated at the same slant range as x.

Therefore, with constant relative measurement precision, each residual has an influence on the fitted curve equal to the proportion of the measured value that the residual represents, regardless of whether the measured value is a large one or a small one. That seems to be a reasonable approach to giving measurements at greater distances an "equal say" in determining the fitted value.

In addition, inasmuch as each weight has a second factor that is inversely proportional to the square of its estimated coefficient of variation,

$$c.\hat{v}.^2 = \frac{\hat{\sigma}^2}{\hat{\mu}^2} = \frac{y^2}{x^2} \quad (8)$$

residuals for measurements of a given mean size are weighted in inverse proportion to their estimated variances. The approach has assumptions and limitations, as does any other statistical method, but it is thought to be reasonably straightforward and useful in these circumstances.

Function Used for This Analysis

The function that was chosen for fitting the neutron measurements was

$$A(r)r^2 = A(r_0)e^{-\frac{r-r_0}{\lambda_0+\delta(r-r_0)}} \quad (9)$$

where $A(r)$ is the activation at slant distance r, and r_0 is the slant distance at the hypocenter, that is, r_0 = burst height. Thus, the scaled values (multiplied by the square of slant distance) are being fitted, and three parameters are estimated:

- $A(r_0)$, the scaled activation at ground zero (hypocenter).
- λ_0, a relaxation length at the hypocenter.
- δ, a change in relaxation length per unit slant distance.

The function $A(r)r^2$ is consistent with the transport dynamics of the neutron energy groups because the rate of change in the exponent decreases gradually with increasing distance. The function gives a very good fit to the DS86-calculated values, and is flexible enough to accommodate a complete range of curves with different rates of change in relaxation length. But it is simple and requires fitting only one parameter in addition to the slope and intercept of the exponential. Some early attempts were made to fit an equation containing a term for $1/r$ in addition to $1/r^2$, but they did not give meaningful results. The effect is too small in comparison with the exponential attenuation to be quantified, especially at the distances for which measurements are made.

Distance Uncertainty

As noted above, the uncertainty in distance, which is the independent variable, is not negligible. A complete and consistent set of distance-uncertainty estimates is not yet available; furthermore, it does not appear feasible to use a statistical method for the fitting that is designed to handle error in the independent variable. In the case of ^{152}Eu, a small subset of four measurements have unusually large distance uncertainties, as estimated by the investigator—50–90 m, compared with less than 30 m for all others. The effect on the fitted curve of removing these measurements was evaluated and found to be negligible. As noted above, RERF is making continuing efforts to reduce this source of error.

Correlated Errors

In producing fitted curves, it would be desirable to take account of all covariance among measurements, especially that known to exist among measurements that share factor or factors, such as a measured value of stable cobalt or europium, a calibration factor for counting efficiency, or a calibration factor for the assay of stable cobalt or europium.

But, it does not appear feasible to identify such sets of measurements exhaustively; it would require minutely detailed information from investigators. However, a major portion of this correlation can be addressed by considering situations that involve multiple measurements by the same investigator at the same site and same distance. First, a weighted mean is calculated for the measurements in the set by using the uncertainty estimates for quantities that are *not* the same for all measurements in the set, whose identity can normally be established with confidence. Then, the propagation-of-error formulas are used successively to estimate the uncertainty estimate in the weighted mean. Finally, the uncertainty in the weighted mean is combined with the uncertainties in shared factors.

Example of Use of Weighted Means to Address Correlated Errors Among Measurements from a Particular Site

Suppose that independently distributed random variables have estimated variances

$$X_i, \hat{\sigma}_{X_i}; Y_i, \hat{\sigma}_{Y_i}; U_i, \hat{\sigma}_{U_i}; V_i, \hat{\sigma}_{V_i},$$

with the index identifying the values for the ith measurement. For instance, X might be the radiation count rate of a sample, Y its stable element assay result, and U and V the calibration factors for those two quantities, respectively (U might be the inverse of the calibrated counting efficiency). Those variables would then combine as factors in a formula for the reported result, specific activity $= SA$, such as

$$SA_i = \frac{X_i U_i}{Y_i V_i} \qquad (10)$$

If one or more of these variables takes on identical values for all measurements in the set, as arises when the same calibration factor is used for more than one measured value—say

$$U_i = u, \hat{\sigma}_{U_i} = \sigma_U, V_i = v, \hat{\sigma}_{V_i} = \sigma_V, \ p \le i \le q$$

—then a weighted mean for the set of measurements from p through q may be calculated for the quantity corresponding to the uncalibrated specific activities, which is designated as follows.

$$\overline{\left(\frac{X}{Y}\right)}$$

The estimated variance of each individual quotient X_i/Y_i is established by application of the quotient rule as

$$\frac{\hat{\text{var}}\left(\dfrac{X_i}{Y_i}\right)}{\left(\dfrac{X_i}{Y_i}\right)^2} = \frac{\hat{\sigma}_{X_i}^2}{X_i^2} + \frac{\hat{\sigma}_{Y_i}^2}{Y_i^2} \Rightarrow \hat{\text{var}}\left(\frac{X_i}{Y_i}\right) = \frac{\hat{\sigma}_{X_i}^2}{y_i^2} + \frac{\hat{\sigma}_{Y_i}^2}{y_i^2} \cdot \left(\frac{x_i}{y_i}\right)^2 \equiv \frac{1}{w_i} \qquad (11)$$

where n the weights w_i are the inverses of the estimated variances. The variance-weighted mean is

$$\overline{\left(\frac{X}{Y}\right)} = \frac{\displaystyle\sum_{i=p}^{q} w_i \left(\frac{x_i}{y_i}\right)}{\displaystyle\sum_{i=p}^{q} w_i} \qquad (12)$$

and the estimated variance of this weighted mean is

$$\hat{\text{var}}\left\{\overline{\left(\frac{X}{Y}\right)}\right\} = \frac{1}{\displaystyle\sum_{i=p}^{q} w_i} \qquad (13)$$

The estimated mean specific activity for the set of measurements is thus

$$\hat{SA} = \overline{\left(\frac{X}{Y}\right)} \cdot \left(\frac{u}{v}\right) = \frac{\sum_{i=p}^{q} w_i \left(\frac{x_i}{y_i}\right)}{\sum_{i=p}^{q} w_i} \cdot \left(\frac{u}{v}\right) \tag{14}$$

and its variance is given by

$$\hat{var}\left(\hat{SA}\right) = \left[\frac{\dfrac{1}{\sum_{i=p}^{q} w_i}}{\left(\dfrac{X}{Y}\right)^2} + \frac{\sigma_U^2}{u^2} + \frac{\sigma_V^2}{v^2}\right] \cdot \overline{\left(\frac{X}{Y}\right)}^2 \cdot \frac{u^2}{v^2}$$

$$= \left\{\frac{u^2}{v^2 \sum_{i=p}^{q} w_i}\right\} + \left\{\sigma_U^2 \cdot \overline{\left(\frac{X}{Y}\right)}^2 \cdot \frac{1}{v^2}\right\} + \left\{\sigma_V^2 \cdot \overline{\left(\frac{X}{Y}\right)}^2 \cdot \frac{u^2}{v^4}\right\} \tag{15}$$

It appears that some minor variation of this formula is sufficient for all the radiometric results, in that the investigators' methods are such that subsets of reported values at the same site share either the first, the second and third, or all three of these:

- The calibration factor for counting efficiency.
- The calibration factor for the assay of stable cobalt or europium.
- The measured value of stable cobalt or europium.

In this analysis, the actual values of X_i, Y_i, U_i, and V_i are typically not known, because the raw data are not available. Rather, the values of only $(X_iU_i)/(Y_iV_i)$ or of $(X_iU_i)/(Y_iV_i)$ and Y_iV_i are known from the source document in which the measurement was reported. Correspondingly, the uncertainty estimates for X, Y, U, and V are typically known or estimated only as fractional standard deviations rather than as raw standard deviations. It is possible to make the necessary calculations with those quantities alone, that is the investigator's reported result,

$$\frac{x_i u}{y_i v}$$

and the squares of the related estimated fractional standard deviations,

$$\frac{\hat{\sigma}_{X_i}^2}{X_i^2} \qquad \frac{\hat{\sigma}_{Y_i}^2}{Y_i^2} \qquad \frac{\hat{\sigma}_{U_i}^2}{U_i^2} \qquad \frac{\hat{\sigma}_{V_i}^2}{V_i^2},$$

by using some simple algebraic identities:

$$\frac{1}{w_i} = \left(\frac{\hat{\sigma}_{X_i}^2}{X_i^2} + \frac{\hat{\sigma}_{Y_i}^2}{Y_i^2} \right) \cdot \left(\frac{x_i}{y_i} \right)^2 \tag{16}$$

$$\hat{SA} = \left(\overline{\frac{X}{Y}} \right) \cdot \left(\frac{u}{v} \right) = \frac{\sum\limits_{i=p}^{q} w_i \left(\frac{x_i}{y_i} \right)}{\sum\limits_{i=p}^{q} w_i} \cdot \left(\frac{u}{v} \right) = \frac{\sum\limits_{i=p}^{q} \dfrac{1}{\left(\dfrac{\hat{\sigma}_{X_i}^2}{X_i^2} + \dfrac{\hat{\sigma}_{Y_i}^2}{Y_i^2} \right) \cdot \left(\dfrac{x_i}{y_i} \right)^2} \cdot \left(\dfrac{x_i}{y_i} \right)}{\sum\limits_{i=p}^{q} \dfrac{1}{\left(\dfrac{\hat{\sigma}_{X_i}^2}{X_i^2} + \dfrac{\hat{\sigma}_{Y_i}^2}{Y_i^2} \right) \cdot \left(\dfrac{x_i}{y_i} \right)^2}} \cdot \left(\frac{u}{v} \right),$$

$$= \frac{\sum\limits_{i=p}^{q} \dfrac{1}{\left(\dfrac{\hat{\sigma}_{X_i}^2}{X_i^2} + \dfrac{\hat{\sigma}_{Y_i}^2}{Y_i^2} \right) \cdot \left(\dfrac{x_i}{y_i} \right) \cdot \left(\dfrac{u}{v} \right)}}{\sum\limits_{i=p}^{q} \dfrac{1}{\left(\dfrac{\hat{\sigma}_{X_i}^2}{X_i^2} + \dfrac{\hat{\sigma}_{Y_i}^2}{Y_i^2} \right) \cdot \left[\left(\dfrac{x_i}{y_i} \right) \cdot \left(\dfrac{u}{v} \right) \right]^2}} \tag{17}$$

$$= \frac{1}{\sum\limits_{i=p}^{q} \dfrac{1}{\left(\dfrac{\hat{\sigma}_{X_i}^2}{X_i^2} + \dfrac{\hat{\sigma}_{Y_i}^2}{Y_i^2} \right) \cdot \left(\dfrac{x_i}{y_i} \right)^2 \cdot \dfrac{u^2}{v^2}}} + \left[\frac{\sigma_U^2}{U^2} + \frac{\sigma_V^2}{V^2} \right] \cdot \left(\frac{X}{Y} \right)^2 \cdot \frac{u^2}{v^2}$$

$$= \frac{1}{\sum\limits_{i=p}^{q} \dfrac{1}{\left(\dfrac{\hat{\sigma}_{X_i}^2}{X_i^2} + \dfrac{\hat{\sigma}_{Y_i}^2}{Y_i^2} \right) \cdot \left(\dfrac{x_i}{y_i} \right)^2 \cdot \dfrac{u^2}{v^2}}} + \left[\frac{\sigma_U^2}{U^2} + \frac{\sigma_V^2}{V^2} \right] \cdot \hat{SA}^2 \tag{18}$$

Apart from that approach of calculating weighted means within sites, some additional analysis might be possible via separate fitting of values for different investigators in cases that involve sufficient numbers of measurements for each investigator. That would serve to quantify any overall systematic difference between investigators that would be due to any combination of the factors above.

Application to Measurements by Radiometric Methods (^{60}Co and ^{152}Eu)

Radiometric methods quantify the neutron activation product nuclide, ^{60}Co or ^{152}Eu, in terms of its radioactive emissions by counting gamma or beta emissions in the sample per unit time. The number of emissions can be corrected by subtracting a background to obtain a net count rate and is then divided by the applicable counting efficiency to obtain a number of atomic disintegrations of the radio-nuclide in the sample per unit time. That number is divided by the amount (mass) of the target element of the neutron reaction in question that is in the sample, which is determined by a separate measurement, to give a specific activity stated in units of radioactivity per unit mass of the stable element. The result is then backcorrected for decay to the time of the bombing by using the radioactive half-life of the acti-vation product. The equation used for calculating specific activity in the sample ATB is thus

$$A_m(ATB) = \frac{X_{net}(ATM) \cdot 2^{\frac{ATM-ATB}{HL}}}{m \cdot m_{calib} \cdot eff} \qquad (19)$$

In Equation 19, ATB is the date of bombing (August 6 or 9, 1945), ATM is the date of the measurement, and the difference $(ATM - ATB)$ is expressed in years. X_{net} is the net count rate for the sample in counts s^{-1}. The mass-related value m is the raw result of the assay of the mass of cobalt or europium in the sample in such units as count rate of an activation radionuclide after neutron irradiation or units of absorption in atomic absorption spectroscopy, and m_{calib} is a calibration factor for that assay; the product of m and m_{calib} is expressed in milligrams. Finally, eff is the applicable counting efficiency in counts $s^{-1} Bq^{-1}$, and HL is the half-life of ^{60}Co (5.2714 yr) or ^{152}Eu (13.54 yr).

All source documents contain an estimate for A_m (ATB) or the necessary infor-mation to calculate it straightforwardly, and every publication used in this analysis contains an estimated standard deviation for each such measurement. In the case of the ^{60}Co results of Hashizume and others (1967), results are reported as count per milligram of cobalt, and an applicable counting efficiency is given. The date of mea-surement has been checked (Maruyama 2000) and can be stated with good confi-dence to be within calendar year 1965, that is, with an uncertainty equating to about 5% standard deviation in the decay factor. Except for the work of Kerr and others (1990) as reported in ORNL 6590, every indication is that the investigators' esti-mated standard deviations were based only on the counting statistics involved in the measurement of $X_{net}(ATM)$, although other uncertainty estimates were usually pub-lished separately for the stable-element assay involved in determining m.

Calibration of Counting Efficiency

With the exception of ORNL 6590, there is no indication that the uncertainty in the calibration of the counting efficiency eff is included in any of the investigators' estimates. This uncertainty can typically be kept low and is not a major concern among investigators. However, it will be estimated here by using typical values for radiation standard sources and source-detector geometry considerations as detailed below, and it has been added to the estimated total uncertainty calculated herein.

Background and Net Count Rate

A more worrisome issue in these circumstances, particularly because of the significance attributed to measurements at low levels, is the determination of background. In all cases, background has been determined with an empty counting chamber (counting-system background) or has been simultaneously estimated from the spectrum of the sample itself by means of a trapezoidal approximation based on the count rates in channels adjacent to the region of interest for a given photon energy range. Counting of system background was generally not determined with prepared blank samples intended to simulate the radiation-scattering properties of the sample material, but this is not a major concern; in fact, it would have been difficult to simulate the radiation-scattering properties of the samples with materials assured not to contain any detectable sources of radiation in the energy region(s) of interest.

A more serious concern is related to true environmental background samples. These would be samples that are assured to contain, as exactly as possible, the same quantities of ^{60}Co and ^{152}Eu and any potential interfering radionuclides as would exist in the bomb fluence samples, from sources other than activation by the bomb neutrons, including activation by cosmic-ray-generated neutrons. Few samples for either ^{60}Co or ^{152}Eu were at distances sufficient to assure no measurable bomb fluence but made of materials otherwise essentially identical to bomb-fluence samples. The few samples that meet these criteria had poor recovery of stable cobalt or europium, and correspondingly lacked the sensitivity to give a quantitative estimate of true environmental background at the levels that are believed to be attributable to production by cosmic-ray neutrons. As shown in Figures B-1 and B-2, the minimum detectable concentrations in the only apparently suitable samples, which were measured by Shizuma and others (1993, 1998) in the cases of both ^{60}Co and ^{152}Eu, are not at least 1–2 powers of 10 below the lowest reported bomb-fluence sample result, as one would wish them to be. Dr. Shizuma has supplied the spectra of several samples for each nuclide, which he regards as being background samples but for which reported values have not been published. Most of the samples were close enough to the hypocenter for some detectable activation to be normally expected, and it is not clear whether there is sufficiently strong independent information to establish that they were not exposed to the bomb fluence (Shizuma 2000b). A thorough analysis should include strong assurance to establish that these samples do not represent part of the statistical variation that should be included in the range of results from exposed samples.

The best current estimate of cosmic-ray-induced background comes from assays of laboratory reagents (stock chemical compounds). The reagents do not necessarily have the same exposure circumstances as the bomb samples in relation to cosmic rays. Bomb-fluence samples were exposed on the surfaces of various human structures from the time of their removal from the earth—with the possible exception of a sample of granite from a large rock at Shirakami Shrine measured by Shizuma and others (1997), before 1945—until they were removed to a storage area at the time of sample collection. Laboratory reagents were removed from the earth at some point and were then in various storage situations. The storage areas of both stored bomb-fluence samples and laboratory reagents might or might not have been well shielded from cosmic rays. Although cosmic rays are very penetrating, neutrons generated by cosmic rays in the atmosphere have an attenuation length of about 10 cm or less in rock, and comparable or greater total overlying thicknesses of rock and concrete are certainly present in the deeper portions of some structures. The dates of removal of bomb-fluence specimens from their in situ locations are known in some cases and not in others. As for laboratory reagents, the documentation of their age is not thorough enough to provide firm evidence that they were at saturated levels in their storage location. For all those reasons, considerable uncertainty exists in the comparative saturation levels of bomb-fluence samples and laboratory reagents, but the former are likely to be more saturated overall. A "best currently available" estimate of cosmic-ray background, corrected for decay to reflect investigators' decay correction to ATB, was subtracted from the measured values for plotting and fitting herein.

These issues are addressed in greater detail below.

Calibration of the Stable Element Assay

As in the case of the assay of sample ^{60}Co and ^{152}Eu, there is no indication that the accuracy in the calibration of the assay of stable cobalt or europium in samples was included by investigators in their estimation of uncertainty in the result of that assay. Again, the estimates appear to have been based only on the counting statistics of neutron activation assays after irradiation in reactors or by other neutron sources, or they were based on similar measures of reproducibility without regard to the accuracy of calibration.

Issues Related to Experimental Design and Data-Quality Assurance

Generally, the assays were well designed to improve sensitivity and reduce the uncertainties in counting statistics, however, their rigor in regard to metrological and administrative issues is unclear.

• Provenance and storage history of samples, including specific controls to provide assurance of proper labeling and identification and detailed histories of storage locations and conditions, are not well documented, largely because of the practical and social difficulties of obtaining samples.
• Traceability and guaranteed accuracy of metrological standards seem reasonably well assured in most cases for radioactivity content but much less so for stable-element content.
• There has been little or no blinding of measurements or other controls that might be used to ensure the integrity of experimental design at various levels, that

is, to ensure that the results used for analysis are statistically representative of all results that could have been obtained from the total pool of available samples available to individual investigators or to all investigators combined.

• There is an absence of quality-assurance procedures of various types that could be used to detect extraneous contamination or cross-contamination of samples with respect to either the radioactive nuclide or the stable target element (for example, prepared blanks in every run). There is no indication of checks on accuracy at low levels of sample activity (such as prepared low-level standards of known activity).

• Some interlaboratory comparisons have been done, but there was no systematic and comprehensive approach to this issue.

Those observations are based on retrospective evaluation of the entire historical body of available measurements in the context of data-quality standards that have been promulgated in recent years for government and commercial laboratories, which process large numbers of samples according to well-developed methods with commensurate resources. The measurements under analysis here have been performed over some 40 years and necessarily at a research stage of development. They have developed in response to rapidly changing questions and perceived needs that arose as the complex process of modeling and measuring the bomb fluences unfolded while computing and measuring technologies were advancing. Measurements have been given limited funding, which is awarded to individual investigators. In the United States only two funded investigators have been performing measurements since DS86. In Japan, after DS86, a dosimetry group based primarily at Hiroshima University has exerted considerable coordination and control, but not all measurers are in this group, and measurements have not been formally subject to some of the kinds of controls described above. It is not possible to assess the impact of these considerations in the present analysis, but Chapter 3 presents recommendations for future work.

REVISED UNCERTAINTY ESTIMATES (^{60}Co AND ^{152}Eu)

Background and Net Count Rate

As noted above, it was not possible to check investigators' estimates of uncertainty based on counting statistics, because of the inability to obtain the raw data. One aspect of this issue can be checked in the limiting case by using a different approach: the estimation of minimum detection limits, which will be discussed below.

Measurements have been corrected by subtracting cosmic background, estimated on the basis of measurements in laboratory reagents and other knowledge documented in Appendix C, and back-corrected for decay from the time of the bombing to the time of the measurement, estimated as being 2 years before the year of publication. In addition, the estimates have been given a fairly large uncertainty (such as

25%) because knowledge of such measures as saturation level was sparse, and that uncertainty has been factored into the total estimated uncertainty of the measurement.

Decay Corrections

There is a basis for correcting a small error in the decay calculation for all ^{152}Eu measurements except those published after 1998 by using a new estimate of the half-life, namely, 13.54 y instead of 13.2 y or 13.3 y, as noted by Iimoto (1999).

The investigators' estimates must be used for the counting error involved in X_{net}, and an error in the decay factor to ATB will be assumed insignificant apart from the correction just noted.

Calibration of Counting Efficiency

The best approach to estimating the error in the calibration of counting efficiency *eff* is to break it down into two components:

• The error in the radioactivity content of the calibration standard as stated by its supplier and the error in the measured value obtained for that standard material with the investigator's equipment.
• The differences caused by differences in the geometry of the bomb-fluence samples and the calibration source.

Those errors are small because of the typically good accuracy and large Bq content of calibration standards in the first instance and because of the care taken by the investigators to devise standards geometrically similar to the bomb samples in the second. On the basis of certificates of analysis and standards of practice, a value of 2% is being used universally for the first factor, pending additional information. A series of experiments reported by Shizuma and others (1993) is extremely helpful in regard to the second factor. A range of error of 1–4% is being used for almost all gamma measurements on the basis of plots in Shizuma and others (1993) and considering the solid angle subtended by the detector, the mass of the sample, and the photon energy being counted.

An exception is the measurements by Okumura and Shimasaki (1997) and Tatsumi-Miyajima (1991). In these cases, a somewhat larger error of 7% has been assumed, pending additional information. It is based on the detector geometry $(< 2\,\pi)$; the fact that a point source in different source-detector positions was used to simulate the unenriched, intact slices of rock being counted, at least in the work by Tatsumi-Miyajima. It is unclear whether or not investigators measured the sample in different, well-defined positions (such as flipped 180°) to correct for possible inhomogeneities in the bomb-induced radioactivity concentration in different parts of the sample.

Assay of Stable Cobalt or Europium

Most investigators give an estimated uncertainty for the assayed sample content of cobalt or europium, usually stated in percentage by weight or parts per million, rather than weight of cobalt or europium in the sample. In some cases, this uncertainty was estimated by a conventional calculation of the sample standard deviation "s" for a set of replicates. In other cases, such as assay by neutron activation analysis (NAA), the uncertainty was apparently based on the Poisson counting statistics of the radiation from the activation-product nuclide in the irradiated sample. In most cases, it appears that, regardless of which approach was used, the investigator treated the calibration of the assay as a known constant, so the estimated error does not include the error in the cobalt of europium content of the standard solution as given by the supplier, nor the error in the calibration result for that material as measured by the investigator's equipment. For example, a 1000 ppm standard solution of europium may contain 1000 ± 30 ppm at the $1 - \sigma$ error level as specified by the supplier, and the irradiation of m_{cal} µg of this solution in a reactor can produce $x \pm \sqrt{x}$ counts in a given counting interval.

Therefore, for the error in determining the mass m of ^{60}Co or ^{152}Eu in the sample, all the errors stated by investigators were used, but an additional uncertainty in the calibration of the assay was estimated as described in detail below.

Cobalt was generally determined by atomic absorption (AA), although measurements performed in the 1960s used a colorimetric assay (based on o-nitrosoresorcin monomethyl ether salt) on samples that were highly concentrated in cobalt by electroplating techniques. Concentrations in original sample material cover a wide range, and concentrations in measured samples an even wider range, because chemical enrichment was used in some cases and not in others. Most of the steel samples measured apparently had cobalt concentrations of about 100 ppm to 300 ppm, but a few were about one-tenth as great. Andesite rock has cobalt at about 10–20 ppm (US Geological Survey [USGS] Certificate of Analysis for andesite, AGV-2), and Nakanishi measured about 20 ppm in roof tile (Nakanishi and others 1983), but he measured only about 0.5 ppm in granite and about 5 ppm in concrete (Roesch 1987). The actual milligram amounts of cobalt in samples at greater distances are given in Table B-3 and reflect chemical enrichment in most cases.

There are some reasons to question whether substantial inhomogeneities in stable cobalt concentration could have existed in some of the steel samples. If inhomogeneities in cobalt concentration existed in samples along with positional dependence of neutron fluence, a quotient of the averaged values of becquerels per gram and milligrams per gram obtained from a homogenized extract might not be representative. Kerr and others (1990) obtained such disparate results from a handrail of a smokestack of the Chugoku Electric Co. that they concluded that it was made primarily from scrap metal. Shizuma and others (1992) obtained very different results for the same samples in succeeding runs using the same procedure: 21.4 vs. 9.96 mg g^{-1} for a steel plate at the A-bomb Dome, 41.2 vs. 11.7 mg g^{-1} for a steel pipe at the Red

Cross Hospital, 37.0 vs. 12.4 mg g^{-1} for a steel ladder at the Red Cross hospital, and 64.6 vs. 15.1 mg g^{-1} for a steel ladder at the Hiroshima Bank of Credit (Shizuma 1997; Shizuma and others 1998). In the case of Shizuma and others (1992), there appears to be a pattern in the recovered concentrations of cobalt, but not a similar pattern in the calculated radioactivity concentration. This would tend to support a batch difference in enrichment chemistry rather than inhomogeneities in the original sample matrices.

Europium content, in contrast, was universally determined by NAA with either a reactor or a ^{252}Cf source, except for the work of Kato and others (1990), which is discussed in more detail below. In some cases, it was not the emissions from the decay of the ground-state isomer of ^{152}Eu that were measured, but the emissions from isomeric transition of a short-lived metastable state, namely the ^{152}Eu state with a half-life of 9.311 h (Shizuma and others 1993). The complicated neutron activation production and decay schemes of the europium isotopes and their isomers, including ^{154}Eu and ^{152}Eu, and the related potential interferences in their spectra in various energy regions of interest require careful assay. The relevant data are given in Table B-2 below.

The assay of ^{152}Eu must consider carefully the time-dependent effect of the metastable state with 9.3-h half-life for any measurement within several days of irradiation. Furthermore, any assay of ^{152}Eu that uses the 121.78-keV photon must consider very carefully any possible counts in the region of interest from the ^{154}Eu photon at 123.07 keV, which could vary considerably with the energy calibration of the system. All those considerations deserve further review and analysis.

The actual native concentrations of europium in sample materials are less variable than those of cobalt. All the reported values of Nakanishi and others (1991, 1993, 1998), Roesch (1987), Shizuma (1997), and Shizuma and others (1993) are in the range of about 0.3 to 3 ppm for concrete, granite, and roofing tiles alike, and the USGS Certificate of Analysis for andesite, AGV-2, is about 1.5 ppm. However, some samples were measured without enrichment, some were enriched via a single-step process (Nakanishi and others 1991; Shizuma 1997; Shizuma and others 1993), and others were enriched via a sophisticated

TABLE B-2 Properties of Europium Nuclides of Interest in Assay of Stable Europium by NAA

Nuclide	HL	Production by N		Photons of Interest		
^{152}Eu	13.537 y	thermal, fast	~39–49 keV	~122 keV	~344 keV	
^{152}Eu	9.3116 h	thermal, fast	39.52 (21.1%)	121.78 (28.58%)	344.28 (26.5%)	
^{154}Eu	8.593 y	thermal, fast	40.12 (38.3%)	121.78 (7.00%)	344.28 (2.38%)	
			39.52 (7.40%)	123.07 (40.79%)		
			40.12 (13.4%)			

multistep process (Nakanishi and others 1991, 1998). The actual amounts recovered in samples, as for cobalt, are given in Tables 3-2a and 3-2b of Chapter 3.

The uncertainty in sample content of stable cobalt or europium is important and differs so much among source documents with regard to the available information and the nature of the assay that a case-by-case discussion is appropriate here. All noted percentage errors given below are 1-SD estimates.

• Saito (Roesch 1987) and Hashizume and others (1967) measured ^{60}Co in enriched steel samples and assayed stable cobalt by the colorimetric method, and they reported uncertainty estimates that are clearly based on reproducibility among replicate measurements. A reproducibility-based estimate of "< 5%" was given. An additional 3% calibration error for the assay is assumed here.

• Nakanishi did a few early measurements of ^{60}Co in unenriched samples of a roofing tile (Nakanishi and others 1983) and in the granite and concrete core of the Fukoku Seimei building (Roesch 1987), which were assayed for cobalt by reactor-based NAA. On the basis of recommended value of ±6.25% in the current USGS Certificate of Analysis for andesite (AGV-2), a total calibration error of 7.5% has been assigned to these results.

• Hoshi and Kato (1987) measured ^{60}Co in unenriched steel samples but do not describe the assay for stable cobalt. A total error of 5% is assumed for stable cobalt, pending further information.

• Kerr and others (1990) used a unique method to measure ^{60}Co in large, intact steel samples using large-area detectors and cross-calibration done by counting a sample before and after enrichment in the large-area detectors and a well detector, respectively. For cobalt in the intact samples, they give extensive documentation of comparisons with samples supplied by the US National Bureau of Standards (now the National Institute of Standards and Technology [NIST]), and they measured cobalt content by several methods. Their estimate of about 3% overall error in stable cobalt content is included in their error calculations for total measurement error.

• Kimura (1993) and Kimura and others (1990) measured ^{60}Co in enriched steel samples. In their earlier paper, they state that stable cobalt was measured by NAA. In the later paper, they state that AA was used for the Yokogawa Bridge sample and NAA for the A-bomb Dome sample. A calibration error of 3% is assumed in addition to the error given by the authors. For the A-bomb Dome sample, because no error is given, a total error of 5% is assumed for stable cobalt.

• Shizuma (1999), and Shizuma and others 1992, 1993, 1998) measured ^{60}Co in unenriched (A-bomb Dome) and enriched steel samples. All stable cobalt assays were done via AA by a commercial laboratory. An error equating to 7.7% is noted for the A-bomb Dome samples, which were unenriched, and an error of 5% is quoted for all the other samples. Given that these estimates are quoted from the results of a commercial laboratory, it is assumed that they include calibration error.

• Okumura and Shimasaki (1997) measured ^{60}Co in unenriched samples of andesite rock and assayed stable cobalt by NAA. Because no error estimate is given by the authors for stable cobalt and because the samples are unenriched, and considering the substantial uncertainty in the values recommended by USGS for the AGV-2 andesite standard (6.25% SD), a total error of 7.5% is assumed for stable cobalt, pending further information.

• Nakanishi and others (1983) and Roesch (1987) originally measured [152]Eu in unenriched granite, tile, and concrete and later in singly and doubly enriched samples (Nakanishi and others 1991, 1998). Assay of stable europium in unenriched samples was said to have been calibrated with geochemical standard rocks from US and Japanese geological-survey agencies (Nakanishi and others 1983). On the basis of current USGS certificate of analysis for andesite (AGV-2), which gives an error of 6.5% for europium, a calibration error of 7.5% is assumed for the unenriched samples. For enriched samples, 1000 ppm europium solution from a chemical supplier, intended for use in AA assays, was used for calibration. A contemporary certificate of analysis from an international supplier, Sigma-Aldrich, gives a range of $\pm 3\%$ for this type of product. The 1991 paper unfortunately does not give error estimates for stable europium. Pending further information, reproducibility and total calibration errors are assumed at 5% and 3%, respectively for this particular publication. For the 1999 paper, a calibration error of 3% is added to the stated error for stable europium.

• Hoshi and Kato (1987) measured [152]Eu in unenriched granite samples, and Hoshi and others (1989) measured [152]Eu in unenriched samples of granite, tile, and concrete. Hoshi and others (1989) give a range of 4–10% for the estimated error in stable europium content, but no estimate is given in Hoshi and Kato (1987). Based on the quoted range and the discussion in Hasai and others (1987), a total uncertainty of 10%, including calibration error, is assumed for all of these measurements, pending additional information.

• Kato and others (1990) measured [152]Eu in enriched granite samples. Stable europium was determined by an electrothermal AA technique (ET-AAS). A calibration error of 3% is assumed to apply in addition to the error estimates given by the authors for stable europium.

• Shizuma and others (1993, 1998) measured [152]Eu in unenriched and enriched granite samples. Stable europium was measured by neutron activation in reactors or with the Geniken [252]Cf source and evaluated on the basis of ratios of paired samples to which 50 µg of europium had and had not been added from a standard solution. It is not yet known how the uncertainty calculation was applied to the ratios, and the accuracy and traceability of the standard solution are unknown. Pending additional information, a calibration error of 5% is used here in addition to the error estimates given by the authors for stable europium.

• Okumura and Shimasaki (1997) and Tatsumi-Miyajima (1991) measured [152]Eu in unenriched samples of andesite rock. Stable europium was assayed by reactor-based NAA, but error estimates are not given. Because the samples were unenriched and because of the error suggested by USGS for determinations in andesite by similar methods, a total error of 7.5% is assumed for stable europium in all these measurements.

Combining Estimates of Uncertainty

In summary, errors for calibration were estimated for this analysis for the counting efficiency and the assay of stable cobalt or europium. The calibration error for counting efficiency was estimated in two separate parts, source-detector geometry and other; "other" includes the accuracy of the radioactive solution used as a standard. These errors and the reproducibility-related errors given by the authors for stable cobalt or europium content must be combined with the counting-statistics errors attributed by the authors to count-rate results to obtain a more realistic estimate of total experimental error. All of these errors are in multiplicative factors shown in Equation 6. They were therefore expressed as percentage errors in a spreadsheet and added in quadrature (square root of sum of squares). When weighted means were calculated as described above for measurements at the same location and same distance by the same investigator, the corresponding formulas equivalent to Equations 16 and 17 were used in the spreadsheet.

Results

The fitted values for DS86 calculation and selected subsets of the measurements are given in Table B-3. These estimates do not include any potential error due to cross contamination, sample-selection bias, or failure to account properly for peak interferences. Plots of measurements with their calculated uncertainty estimates depicted as error bars are included in the body of Chapter 3. Details of the RERF database and the uncertainty calculations will be made available on the RERF Web site at rerf.or.jp.

The results shown in Table B-3 include confidence intervals that reflect the actual dispersion in the measured data about the fitted curve, which is considerably greater than would be suggested by the uncertainty estimates for the measurements that are separately calculated by propagation of error in this appendix. Those uncertainty estimates are used as weights in the regression to give more influence to the more precise measurements, but they do not determine the magnitude of the standard errors that the nonlinear least-squares regression routine estimates for the fitted parameters. This total error is estimated by the regression routine based on the properties of the weighted "sum of the squares of residuals" that is being minimized in the least-squares fit.

The results shown in Table B-3 include confidence intervals that reflect the actual dispersion in the measured data about the fitted curve, which is considerably greater than would be suggested by the uncertainty estimates for the measurements that are separately calculated by propagation of the error. For example, simulations performed with the ^{60}Co data suggest that the confidence interval for the relaxation

TABLE B-3 Fitted Neutron Activation Values Using the Total Uncertainty Estimates: Hiroshima

^{60}Co	DS86	All Measurements	Measurements at Slant Range <1000 m	Measurements at Slant Range <1400 m (does not include Yokogawa Bridge)	Measurements at Slant Range <1500 m (includes Yokogawa Bridge)
$A(r_0)^a$, Bq mg^{-1}	18	11 [10,13]	12 [10,15]	12 [10,14]	11 [10,13]
λ_0*, m	119	120 [95,145]	76 [28,124]	93 [63,122]	123 [99,147]
$\delta(1000)$*, m km^{-1}	11	51 [15,88]	170 [49,300]	130 [64,190]	46 [10,81]

^{152}Eu	DS86	All Measurements	Measurements at Slant Range <1200 m, All Investigators	Measurements by Nakanishi et al., All Distances	Measurements by Shizuma et al. and Hoshi et al., All Distances
$A(r_0)^a$, Bq mg-1	169	110 [101,119]	110 [101,119]	113 [95,131]	107 [98,119]
λ_0*, m	119	134 [116,153]	128 [104,152]	137 [102,172]	125 [97,152]
$\delta(1000)$*, m km^{-1}	11	75 [42,110]	92 [31,150]	64 [10,120]	110 [51,170]

[a] $A(r_0)$ is activation at the hypocenter and λ_0 is relaxation length at the hypocenter. $\delta(1000)$ is the change in relaxation length, in m per km slant distance. Units for $\delta(1000)$ are shown as m per km for clarity although this quantity is technically unitless and could be expressed as a fraction or a percent.

NOTE: Values in brackets are approximate 95% confidence limits for the indicated parameter estimates based on asymptotic approximations supplied by the Stata version 6.0 nonlinear least-squares regression procedure "nl."

length at the hypocenter would be less than half as wide as the confidence interval shown in Table B-3. Some of this apparent over-dispersion in the measurements could clearly be reduced by using detailed models of samples and their environs to create more accurate sample-specific calculated values for all of the measurements. Some of this over-dispersion, however, might also reflect sources of random error in the measurement process that are still unknown.

Issues Related to Background, Spurious Signal, and Detection Limits

The presence of background is a sine qua non of radiometric measurements and is of particular interest in the type of low-level measurement that is necessary for neutron activation products from the atomic bombs in Hiroshima and Nagasaki. In addition to counting-system intrinsic background and the background levels of the radionuclide being measured that might be present in a sample because of sources other than the atomic bombs, this report touches on potential sources of misleading signal (counts) that might more appropriately be denoted as counting "interferences." Statistically defined limits of detectability are calculated and compared with both the possible naturally occurring background levels and the bomb-induced levels of interest at greater distances from the hypocenter.

Issues related to the detection of ^{36}Cl by accelerator mass spectroscopy are different from those related to the radiometric detection of other thermal-neutron activation products because AMS is a fundamentally different type of measurement process. The available information on it is also different, as explained below.

Quantification of Minimum Detectable Concentrations (MDCs) of Total Sample Radioactivity per Unit Mass of Target Element

The quantification of detection limits is critically important to the elucidation of the background issue. The first question to be asked for a given measurement method is how much <u>total</u> radioactivity in the sample, regardless of how it came to be there, can reliably be distinguished from the counting-system background that applies to nonradioactive samples. This detection limit, which would be based on a count with no sample or a nonradioactive blank sample present in the counting chamber, provides an indication of the level below which *no reliable determination can be made* regarding the presence or absence of *any* radioactivity in the sample, whether it might arise from cosmic background, the bomb fluence, some other unidentified source, or any combination of these factors. The MDC that actually applies to a given sample with respect to bomb fluence is defined by the total background count rate that applies to that sample, including background from cosmic-ray activation and other possible sources.

To obtain some generic values for the analysis of the various radiometric methods, it is assumed that sample and background are counted for equal intervals, that type I and type II error rates are set at 5%, and that the following formula is therefore applicable, as defined in NUREG 1507 (1995):

$$MDC = \frac{3 + 4.65\, s_B}{KT} \ Bq$$

where s_B is the standard deviation of the background count (for a nonradioactive sample) in an interval of T seconds. It is further assumed that K is a counting efficiency that has been estimated with good precision, so that a single value may be used for K in illustrative calculations. (The uncertainty in the value of K is discussed further below and in other parts of this appendix.) A practical limiting value for counting time T is defined on a basis that is stated below and is used for calculating the tabulated values. The assumption of Poisson counting statistics is taken to apply uniformly to the counting data for radioactivity counting, with no sources of extra Poisson variation at the level of replicate counting measurements on identical samples, and s_B may therefore be taken simply as the square root of the total count.

In actual practice, the "background" for a sample is often obtained by a spectral method from the spectrum for the sample itself, by a trapezoidal approximation in which channels near the photon energy of interest are assumed to represent background levels for those adjacent channels. A line is drawn across the channels in which the photon of interest appears, from the channels on one side of the photon energy region of interest to the other, and used to distinguish "peak counts" from "background counts". The nature of this process is such that the uncertainties of the resulting numbers are not estimable in the same simple way as those of separate counting intervals conforming to Poisson statistics. For the present purpose, it is felt that MDCs derived from the assumption of a separate background count and equal counting intervals, which roughly equate to a trapezoidal approximation using equal numbers of channels in the emitted photon and background regions of interest, are still representative.

For two representative amounts of recovered stable element that might be present in the sample, a minimum detectable *specific activity* in the form of a concentration, denoted MDC_{sa}, is then calculated in units of becquerels per milligram of the target element present in the sample on the basis of the measures involved (sample weight, elemental composition, and fraction of the element recovered in an enrichment process). That is, $MDC_{sa} = MDC/wt_{ele}$, where wt_{ele} is the weight in milligrams of the stable target element in the sample based on the parameters involved (sample weight, elemental composition, and fraction of the element recovered in an enrichment process). These representative amounts are chosen to bracket the amounts that were present in the samples on which a particular set of measurements of interest, typically at longer ranges, was reported by the investigator.

Finally, the quantities are "decayed back" to the time of the bombing by using a specific assumed measurement date of August of a stated year that approximates the date when some measurements of that type were made, for purposes of allowing a comparison with the reported values of measured specific activity. For samples whose radioactivity contents are assumed to be bomb-induced, investigators have uniformly reported Hiroshima and Nagasaki data as of August 6 or 9, 1945, or ATB. Often, the calculated quantity is the only value available; the raw counting data are not available. The MDC stated in terms of specific radioactivity per unit mass of the stable target element, as of the time of the bombing (Bq/mg ATB), is denoted $MDC_{sa,ATB}$.

Another important quantity is the critical level for a given counting system and method. This is the level that is used for declaring a given measurement to be distinguishably different from background. As with the MDC, a 5% type I error rate for a one-sided test is assumed, and the formula

$$L_c = 1.645\sqrt{2s_B^2} = 2.33\,s_B$$

is therefore used.

The critical level can also be expressed in terms of Bq/mg ATB for the applicable parameters of a specific measurement. That provides a guideline in the sense that any reported value less than this limit would have a probability greater than 0.05 of having arisen by chance alone in a sample containing only background, on the basis of the critical region for a one-tailed test with H_0: sample net count rate $= 0$. Furthermore, for measurements below this limit, the type I error rate for a one-tailed test would be at least as great, and in some cases considerably greater than 0.05, if the presence of *background plus the DS86-calculated value* is taken as the null hypothesis.

For the situations of interest here,

$$\text{MDC} \cong 2L_c, \quad \text{therefore,} \quad \text{MDC}_{\text{sa,ATB}} \cong 2L_{C_{\text{sa,ATB}}}$$

Technically, observed net values stated as numbers of counts that lie between the L_c and the MDC are "significant" in the sense that they have a probability of less than 0.05 of having arisen by chance alone if there is no radioactivity apart from the defined background level in the sample.

However, the experiment is not sufficiently sensitive to detect, with acceptable reliability, amounts of radioactivity in a sample that correspond to a mean count between the L_c and the MDC. Those amounts do not have at least a 95% probability of giving a count exceeding the L_c (that is, the type I error rate might be marginally acceptable, but the type II error rate, where it is of concern, is *not* acceptable.)

Furthermore, there is additional uncertainty in the calibration of counting efficiency and, more notably, in the assayed value of the sample's content of the target element. These uncertainties affect the size of the $\text{MDC}_{\text{sa,ATB}}$ that corresponds to a given MDC.

For all those reasons, and others discussed below, reported results close to or below the nominal MDCs in Table B-4 should be regarded as requiring great care in their interpretation.

a. The counting-system background for nonradioactive samples and the applicable counting efficiency

i. ^{60}Co

1. Detection of the 1173- and 1332-keV gammas by a well-type HPGe detector (Shizuma and others 1992; Kimura and others 1990)

Shizuma and others use a 54-mm-diameter by 60-mm-long detector whose combined efficiency for the two gamma rays of ^{60}Co is stated as being close to 0.08 cps/Bq in Table 2 of the 1998 paper (Shizuma 1998) for samples of the size (about 1 g) of interest here. Background is estimated by the trapezoidal method. For the purposes of the present calculation, the background count rate that is assumed to apply is that obtained in a 1288030-s measurement of the Army Food Storehouse background sample: $230/1288030 = 0.00018$ cps. This sample is assumed to represent a reasonable approximation of the applicable system background because the "background counts" value was determined by the trapezoidal method and was consistent with the other "background counts" count rates determined by the same method for the other samples.

Kimura and others (1990) used a detector whose efficiency was stated in their paper as being 0.02849 for the 1.33-MeV gamma rays. The combined efficiency for the two gamma rays is therefore assumed to be about 0.06. As a provisional estimate of background for Kimura and others (1990), the value quoted by Kerr and others (1990) in ORNL 6590 is used: about 200 counts per million seconds per 1-keV channel in a 113-cm^2, well-shielded HPGe detector. A 6-keV-wide total region for the two gamma rays is assumed. It does not appear that better information will be available; the necessary records have not been retained.

2. Detection of the 1173- and 1332-keV gammas by a large-area NaI (Tl) detector (Kerr and others 1990)

On the basis of a cross calibration involving the counting of a sample before and after chemical enrichment, the applicable counting efficiency was stated as 3.99 ± 0.24%—that is, 0.040 cps/Bq—for the paired large (300-mm diameter by 200-mm thick) detectors. The calibration method implicitly includes the total self-absorption in a large (for example, 4 kg) intact steel sample. On the basis of the measurement of the Homestake Mine sample, which was used by the investigators as a background sample, the background count rate was 0.000602 cps for detector 9, or 337 counts in an interval of 560460 s, although the rate for the other detector was almost twice that. Those values are assumed to represent good approximations of system background because the sample was heavily shielded from cosmic rays.

3. Detection of the 318-keV E_{max} beta by GM-plastic scintillation coincidence detector (Hashizume and others 1967)

In both the 1967 paper (Hashizume and others 1967) and the review by Maruyama and Kawamura in the DS86 Final Report, Vol. 2, the counting efficiency is given as 12% and the background as 0.069 cpm.

ii. ^{152}Eu

1. Detection of the 39–40-keV Sm rays by a planar HPGe detector (Nakanishi and others 1998).

The counting efficiency is given in Figure 3 of Dr. Nakanishi's most recent paper (1998) and varies from about 0.014 cps/Bq for a 200-mg sample to about 0.008 cps/Bq for an 850-mg sample. For the purposes of the present calculation, the value of 0.008 is used for conservatism. Counting-system background does not seem to be documented in any of the available publications except for a spectrum for a roof-tile specimen for the older planar Ge (Li) detector, as shown in Figure 2 of the 1983 letter to *Nature* (Nakanishi and others 1983). Apparent background numbers for energies near 39–40 keV were read from this plot and used for a crude calculation, pending additional information from Nakanishi.

2. Detection of the 39–40 keV Sm rays by a coaxial SiLi detector (Iimoto 1999).

The counting efficiency is documented in the paper as 0.002938 cps/Bq. On the basis of a background-sample spectrum provided by Iimoto, the background count rate is about 12 counts/channel in 1.8×10^6 s, or about 227 counts per million seconds in the noted 34-channel region of interest.

3. Detection of the 122 and 344 keV gammas by a well-type HPGe detector (Shizuma and others 1993, 1997).

The counting efficiency is documented in the 1993 *Health Physics* paper (Figure 5 and related text by Shizuma and others 1998) as being fairly constant at about 0.05 cps/Bq for the 122-keV gamma and 0.04 cps/Bq for the 344-keV gamma, over the range of applicable sample sizes. Background count rates are discussed extensively in the 1992 paper in *Nuclear Instruments and Methods* (NIM B66 459-464) and are summarized in Table 4 of that paper. The anti-coincidence values for the well-type detector were determined by the trapezoidal method from a sample containing appreciable ^{152}Eu have been adopted here for the present calculation.

b. Available per-sample amounts of the stable element after enrichment
i. ^{60}Co

The recovered amounts of cobalt for the relevant measurements of Shizuma and others (1998) are well described in their paper in *Health Physics*. They can be calculated unequivocally from Table 1 and range from 2.2 mg for the 4571-m background sample, which was unfortunately the lowest amount of all the samples, to 73 mg for one of the 1481-m samples. Apart from the background sample at 4571 m, however, each of the longer ranges with a measurement of interest (1014 m, 1481 m, 1484 m, and 1703 m) had at least one sample with at least 26 mg of cobalt.

Kimura and others (1990) report values of 239–314 mg for their samples. The former number was calculated from other values given in the 1990 paper and confirmed by writing to the authors; the latter is given explicitly in the 1993 paper (Kimura 1993).

Kerr and others (1990) give extensive data in Table 5 of ORNL 6590 for the stable-cobalt content of the large (about 4 kg) samples that they measured intact in the large-area detector at PNL. These values range from about 120 mg for the Homestake mine sample, which (as in the case of Shizuma) is the lowest of all the samples, to about 1000 mg.

Hashizume and others (1967) do not give the amounts of recovered stable cobalt in their original 1967 paper. The review by Maruyama and Kuramoto (1987) states that not more than 9 mg was recovered but does not give a minimum or sample-specific values.

ii. ^{152}Eu

Shizuma and others (1993) give extensive sample specific information on ppm concentrations in, for example, the 1993 paper. Multiplication by the apparent sample mass of about 3.5 g for all samples resulted in values of 0.02–0.05 mg per sample.

Nakanishi and others (1991) use a more extensive chemical extraction technique and recover comparatively large amounts of europium. Values for extracted specimens were estimated by multiplying the recovery factors, given on page 75 of the 1991 paper, by the range of in situ values recorded in various references beginning with the 1983 letter to *Nature* (Nakanishi and others 1983). Additional information is found in handouts distributed by Nakanishi at committee meetings.

c. Corresponding MDCs and critical levels

The fundamental MDCs based on counting-system background only are shown in Table B-4. A revision of these MDCs in the case of ^{60}Co to account for natural ^{60}Co from cosmic-ray production, is discussed below:

TABLE B-4 Detection Limits for Thermal Neutron Activation Measurements in Hiroshima and Nagasaki

Radionuclide Investigator(s)	^{60}Co Shizuma	^{60}Co Kimura/ Hamada	^{60}Co Kerr	^{60}Co Hashizume/ Maruyama	^{60}Co Okumura/ Shimazaki
Photon Energy Measured	1173 keV + 1332 keV	1173 keV + 1332 keV	1173 keV + 1332 keV	beta coincidence 0–318 keV (25–250 keV window)	
T, seconds	1.00×10^{-6}	1.00×10^{-6}	1.00×10^{-6}	1.00×10^{-6}	
B, cts on interval of T sec.	1.79E + 02	1200	6.02E + 02	1.15E + 03	
s_B	13.36291	34.64102	24.52889	33.91164992	
counting efficiency, cps/Bq	0.08	0.06	0.04	0.12	
MDC, Bq	8.14×10^{-4}	2.73×10^{-3}	2.93×10^{-3}	1.34×10^{-3}	
lowest mg per sample	2.19	240	120		
highest MDC_{sa}, Bq/mg	3.72×10^{-4}	1.14×10^{-5}	2.44×10^{-5}		
highest mg per sample	88.1	310	1000	9	
lowest MDC_{sa}, Bq/mg	9.24×10^{-6}	8.82×10^{-6}	2.93×10^{-6}	1.49×10^{-4}	
HL	5.2719	5.2719	5.2719	5.2719	
year of measurement	1995	1990	1988	1965	
highest $MDC_{sa,ATB}$, Bq/mg ATB	0.26628	0.00423	0.00696		
lowest $MDC_{sa,ATB}$, Bq/mg ATB	0.00662	0.00327	0.00083	0.00206	
Radionuclide	^{152}Eu	^{152}Eu	^{152}Eu	^{152}Eu	^{152}Eu

(*continued*)

TABLE B-4 *(Continued)*

Radionuclide Investigator(s)	^{60}Co Shizuma	^{60}Co Kimura/ Hamada	^{60}Co Kerr	^{60}Co Hashizume/ Maruyama	^{60}Co Okumura/ Shimazaki
Investigator(s)	Shizuma	Shizuma	Nakanishi	Iimoto	Okumura/ Shimazaki
Photon Energy Measured	122 keV + 344 keV	344 keV	39–40 keV	39–40 keV	
T, seconds	1.00×10^{-6}	1.00×10^{-6}	1.00×10^{-6}	1.00×10^{-6}	
B, cts on interval of T sec.	3.42×10^{-3}	1.25×10^{-3}	560	226.6666667	
s_B	58.45226	35.35534	23.66432	15.05545305	
counting efficiency, cps/Bq	0.09	0.04	0.008	0.003	
MDC, Bq	3.05×10^{-3}	4.19×10^{-3}	1.41×10^{-2}	2.43×10^{-2}	
lowest mg per sample	0.02	0.02	1	1.16	
highest MDC$_{sa}$, Bq/mg	1.53×10^{-1}	2.09×10^{-1}	1.41×10^{-2}	2.10×10^{-2}	
highest mg per sample	0.05	0.05	2	1.91	
lowest MDCs, Bq/mg	6.11×10^{-2}	8.37×10^{-2}	7.06×10^{-3}	1.27×10^{-2}	
HL	13.54	13.54	13.54	13.54	
year of measurement	1992	1992	1992	1997	
highest MDC$_{sa,ATB}$, Bq/mg ATB	1.69310	2.32063	0.15670	0.30053	
lowest MDC$_{sa,ATB}$, Bq/mg ATB	0.67724	0.92825	0.07835	0.18252	

NOTES: 1. The background for Hamada's ^{60}Co measurement is based on Kerr's noted background of about 200 counts per million seconds per 1-keV channel in a 113-cm^2, well-shielded HPGe detector (ORNL 6590) and a 6-keV-wide total ROI for the two gamma rays, pending additional information from Hamada. 2. The background for Nakanishi's ^{152}Eu measurements is based on a crude estimate from Figure 2 of Nakanishi's 1983 paper in *Nature,* giving an apparent background of about 60 counts in 118.35 h = 140 counts per channel per million seconds, and assuming a four-channel-wide ROI, pending additional information from Nakanishi.

^{60}Co

Calculated MDC and critical level values are shown in Figure B-1. A value of 1 million seconds for both sample and background is used here as a nominal counting time for comparing results among investigators. Some investigators count for somewhat longer, in which case the MDC would decrease as the inverse of the square root of the counting time if sample and background counting times were increased equally. However, determining background by trapezoidal approximation from the sample spectrum itself introduces additional uncertainty. And if background is determined by a separate count with an empty sample chamber, there should be some

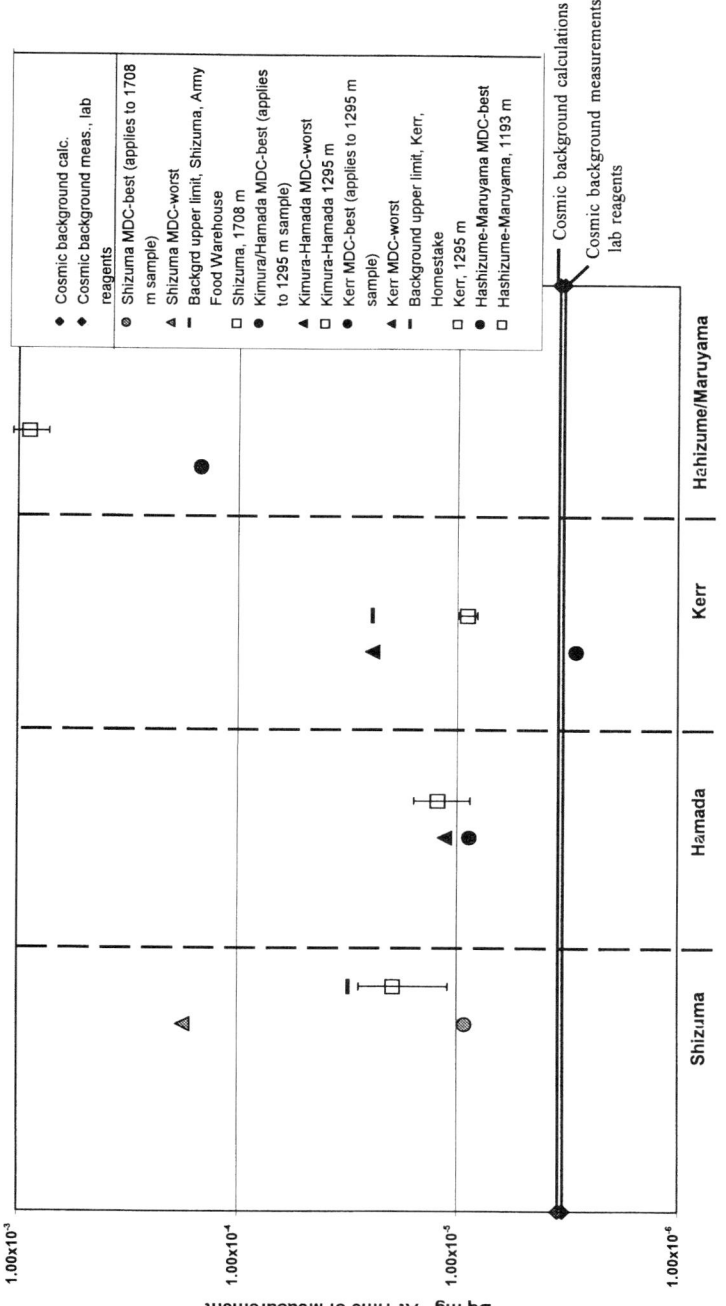

FIGURE B-1 ^{60}Co detection limits.

concern about whether the statistical properties of background count rate were adequately evaluated for such factors as drift in the instrument electronics over long periods and periodic variation due to solar activity. Furthermore, in the case of ^{60}Co counted in recent years in samples from locations far from the hypocenter, there is a nontrivial contribution from cosmic-ray neutrons that increases the total background count rate beyond what is evaluated by either of the two methods above and correspondingly increases the MDC.

ii. ^{152}Eu

Calculated MDC and critical level values are shown in Figure B-2. A value of 1 million seconds is used as a nominal counting time for comparing results among investigators. All the considerations cited above for ^{60}Co also apply to ^{152}Eu except that the cosmic-ray-generated background is not expected to make a significant contribution to experimental error.

iii. ^{36}Cl

AMS results intrinsically report the isotope ratio of ^{36}Cl to chlorine (^{36}Cl/Cl). In this case, there is a system background for a condition of no injected sample that consists of electronic noise in the detectors being used for ^{36}Cl and chlorine, but it is not typically reported. Ratios obtained for blank samples involve a source of stable chlorine of some type, which contains ^{36}Cl at a level defined by the long-term (geological) saturation of the cosmic ray activation in the source material from which the chlorine was taken. Published data on the intrinsic detection limits of the method indicate that it is about 1 atom of ^{36}Cl per 1015 atoms of chlorine (Straume and others 1994). Although that value is not precise or clearly stated from a statistical point of view, it is about one-hundredth of the background levels of interest that appear to exist in unexposed samples. Thus, the situation for ^{36}Cl is different from that of ^{60}Co and ^{152}Eu: the MDC is determined completely by the statistical variation in the background due to cosmic-ray activation.

Calculated Estimates of In Situ Cosmic-ray Production

See Appendix C for a detailed discussion of cosmic-ray neutron fluences.

^{60}Co

Komura and Yousef (1998) give a calculated value of 0.2 dpm/g (3.3×10^{-6} Bq/mg) at saturation, on the basis of a flux of 0.008 n/cm^2 s, in their 1998 abstract of a presentation at the 41st meeting of the Japan Radiation Research Society (December 2–4, 1998, Nagasaki). That value is also cited by Shizuma and others in their 1998 *Health Physics* paper.

^{152}Eu

Komura and Yousef give a calculated value of 5 dpm/g (8.3×10^{-5} Bq/mg), based on a flux of 0.008 n/cm^2/s in the 1998 abstract just cited.

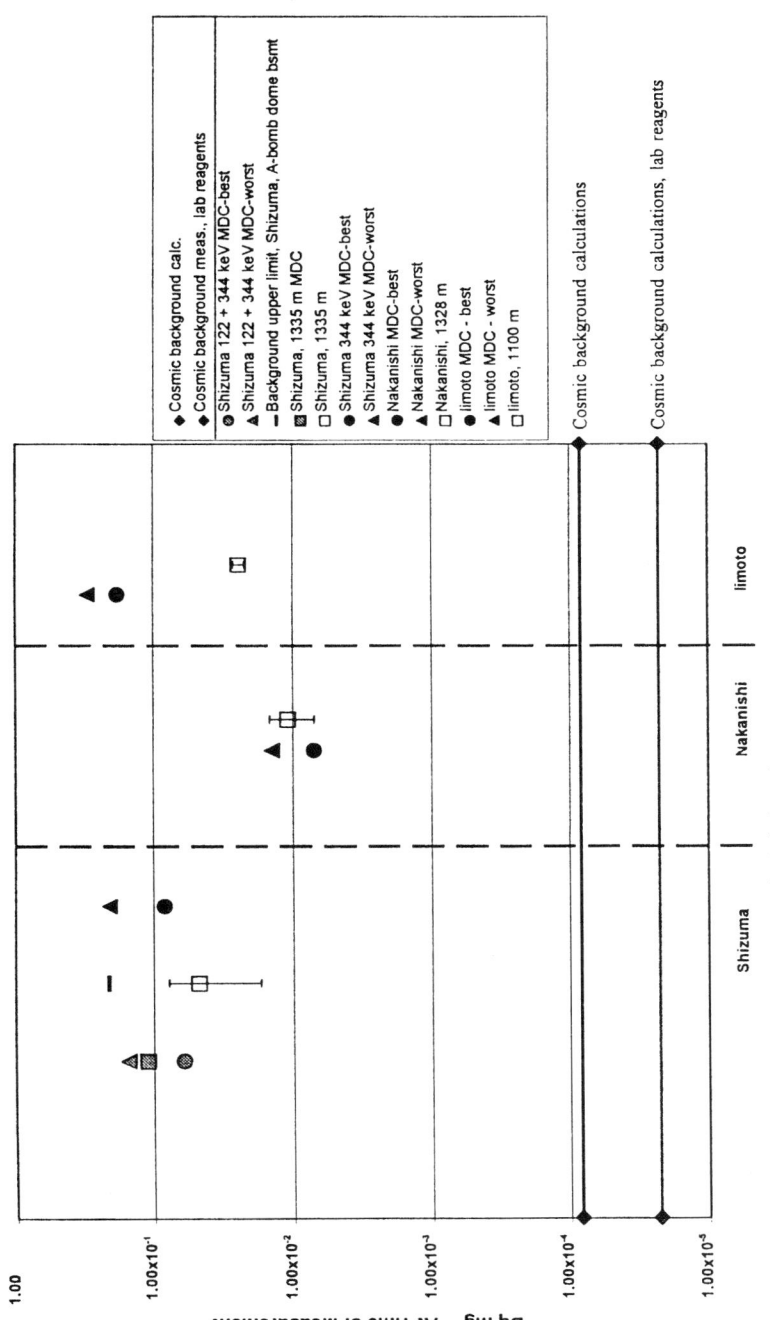

FIGURE B-2 ^{152}Eu detection limits.

Measurements of Samples Far from the Hypocenters
or Heavily Shielded Samples

^{60}Co

Kerr and others (1990) measured two samples, one from a surface building and one from a mine. Both had rather meager cobalt content, and the MDC is about 2.4×10^{-5} Bq/mg as measured for the mine sample and 1.5×10^{-5} Bq/mg for the surface sample. Shizuma and others (1998) measured a sample from the Army Food Warehouse at 4571 m in Hiroshima. It also had little stable cobalt in the extracted sample, and the MDC was about 3.7×10^{-4} Bq/mg.

^{152}Eu

Shizuma and others (1992) measured a heavily shielded sample from the basement of the A-bomb Dome, but its MDC was about 0.21 Bq/mg as measured, not nearly low enough to be informative. Shizuma also supplied some results from the Hiroshima Commercial High School at a 2870-m ground distance (Shizuma 2000a). It appears that this sample also had a poor recovery of 1.82 ppm in an enriched sample of 6.75 g, equaling about 0.0123 mg, resulting in an MDC of about 0.341 Bq/mg at time of measurement in 2000, or about 5 Bq/mg ATB in 1945.

Trends in Deep Portions of Cored Samples

^{60}Co

Some early attempts were made to measure depth profiles in steel, for example, the work on Aioi Bridge girders reported by Hoshi and Kato (1987) and the work by Shizuma and others (1992) on structural steel of the A-bomb Dome, but these measurements are far too shallow to approach the asymptote.

^{152}Eu

A thorough review indicates that a total of six cores or samples of a similar nature have been measured in granite or concrete. Of these, two (the Saikou-ji gravestone and the Motoyasu Bridge pillar) are of such small cross section that the effective depth of the deepest samples is not what is indicated by the axial distance on the depth profile. Of the others, none is measured at a depth greater than 37 cm, and no apparent approach to an asymptote is seen in the depth profiles. Two of the cores measured by Shizuma and others (1997) are deep enough to allow measurement at somewhat greater depths (Hiroshima Bank, 62 cm; Shirakami Shrine; 81 cm), but indicates that those deeper slices would fall below the MDC.

^{36}Cl

On the basis of material presented by Tore Straume at the workshop on RERF dosimetry held on March 13–14, 2000, in Hiroshima, measurements deep in concrete appear to approach an asymptote in the vicinity of 100 Bq mg^{-1} at depths greater than 35 cm. That is consistent with background samples measured in a shielded location at 1700 m and that with other background samples reported by Straume and others (1994).

Measurements of Laboratory Reagents

^{60}Co

Komura and Yousef (1998) measured laboratory reagents containing large amounts of stable cobalt in a subterranean laboratory, giving very low detection limits. They reported a measured value of 0.21 dpm/g (3.5×10^{-6} Bq/mg), comparable with calculated values. Shizuma (1999) also measured ^{60}Co in 4 g of CoO with a well-type Ge detector and obtained a much lower value of 7.2×10^{-7} Bq/mg. It is possible that this sample was not old enough to have reached saturation at background levels in its storage location or was stored in a heavily shielded location.

^{152}Eu

Komura and Yousef (1998) report a value of 1.37 dpm/g (2.3×10^{-5} Bq/mg) in their measurements on Eu_2O_3 that is described as "modern" with respect to age (as opposed to before World War II), which is only about one-fourth of the calculated value. Shizuma (1999) reported an even lower value of 4.6×10^{-6} in 1 g of Eu_2O_3 reported to be about 25 years old. Again, there are major unresolved questions about the saturation levels of both Eu_2O_3 samples.

Potential Counting Interferences

Several potential causes of misleading results have been identified by reviewing the literature and interviewing the investigators. The specific possibilities mentioned here are preliminary and require further investigation. Spectral and radiochemical data were taken from the *WWW Table of Radioactive Isotopes* by Chu, Ekstrom, and Firestone, of the Lawrence Berkeley National Laboratory in the United States and the University of Lund in Sweden at the Internet address http://nucleardata.nuclear.lu.se/nucleardata/toi/index.asp.

Lanthanum X-rays from Photon Interactions
in Stable Lanthanum of Sample Matrix

Nakanishi and others have primarily counted ^{152}Eu via the K_α x-rays of samarium that are emitted at 39.522 keV and 40.118 keV. Because of the chemistry of the europium enrichment process and the natural abundance of lanthanum in the lithosphere, the samples are expected to contain large amounts of lanthanum. Lanthanum has K_β rays at 37.720 keV, 37.801 keV, 38.804 keV, 38.726 keV, and 38.826 keV. All can be produced by interactions of higher-energy photons, such as background photons and photons from ^{152}Eu, with inner-shell electrons in the lanthanum contained in the sample.

^{138}La and ^{176}Lu Relative to ^{152}Eu in the 39 to 40-keV Region

^{138}La and ^{176}Lu are extremely long-lived (over 10^{10} y) naturally occurring lanthanides that might be expected to accompany ^{152}Eu in chemical separations. Their natural isotopic abundances are 0.0902% and 2.59%, respectively. By calculation, natural lanthanum should be about 0.82 Bq/g [138]La/La and natural lutetium should be about 52 Bq/g [176]Lu/Lu. [138]La emits Ba x-rays at 36.304 keV (1.19%), 36.378 keV (3.69%), 37.255 keV (1.16%), and 37.349 keV (0.261%), and five Ce x-rays at 39.17–40.33 keV, all with low spectral abundance under 0.01%. This source of interference is expected to be minor compared with the lanthanum x-rays.

[176]Lu emits no photons at these precise energies but does emit a number of lower—and higher—energy photons and was specifically identified by Nakanishi as a possible concern, perhaps because of its substantial natural abundance.

[212]Pb([212]Bi) Relative to [152]Eu in the 39 to 40-keV Region

[212]Bi of the thorium series (half-life, 60.55 m) has a 39.9-keV gamma with 1.1% spectral abundance associated with its alpha decay.

[223]Ra and [234]U Relative to [152]Eu in the 122 keV Region

[223]Ra (half-life, 11.435 d), a naturally occurring member of the actinium (4n+3) series, has a gamma ray of 1.2% spectral abundance at 122.319 keV. [234]U has a gamma ray of 0.034% spectral abundance at 120.90 keV.

[227]Ac Relative to [152]Eu in the 122-keV Region

[227]Ac (half-life, 21.773 y) of the actinium series, has a gamma ray of low spectral abundance (0.00213%) at 121.53 keV.

Discussion

In many cases, the uncertainty estimates calculated for this analysis are substantially greater than those published by the authors originally. That is due primarily to the inclusion of terms involving the assay of stable cobalt or europium. Estimates of the precision of the stable cobalt or europium assay based on measures of reproducibility have often been included in publications but have almost never been included in a combined estimate of total uncertainty. The accuracy (as opposed to the precision) of the assays is also of concern, especially in cases involving unenriched samples. The calibration of the assays of stable cobalt or europium in unenriched samples appears to have unexpectedly substantial uncertainty, according to the information that has been obtained to date. Thus, the values estimated here for the uncertainty in the reported values of stable cobalt or europium and the total combined uncertainty of the specific radioactivity per unit mass of stable cobalt or europium, are often considerably larger than might have been suggested previously.

The most important effect of these revised estimates, in a proportional sense, is to increase somewhat the unrealistically low uncertainty that was sometimes estimated for measurements with good counting statistics because of their relatively large radioactivity content.

Table B-3 offers some interesting observations. In all cases, the measurements suggest that the DS86-calculated value is (very roughly) 50% too high at the shortest ranges. Interestingly, the initial fitted relaxation lengths are *not* particularly discrepant compared with DS86. But the measurements clearly display substantially greater curvature than DS86, as quantified by the δ parameter for change in relaxation length per unit slant range. Several selected subsets of the data are shown in the table. They serve mainly to illustrate that the trends are not strongly influenced by the less precise data at longer ranges. For ^{60}Co, three subsets show the effect of omitting successively, in order of decreasing range. The highly discrepant measurements at distances beyond the Yokogawa Bridge; the measurements at the Yokogawa Bridge, for which the fully calculated DS86 values involve a large correction from free-field values because of the unusual nature of the structure in which the samples were constituent; the measurements at 1168 and 1330 m that are thought to be above the MDC but are not completely documented with respect to mg cobalt content.

In the case of ^{152}Eu, essentially all measurements beyond 1200 m are suspect with respect to the MDC, at least pending some additional information on several measurements by Nakanishi and others. However, these more distant measurements do not account for the curvature in the fitted values. There is a difference between the fitted values for the two main groups making measurements, the Nakanishi group and the Shizuma group, but the difference does not appear to be statistically significant.

The Nagasaki measurements, in contrast with those in Hiroshima, do not appear to support statistically a discrepancy with DS86, on the basis of the methods described here that have been applied to them thus far. However, the *absence* of a discrepancy in Nagasaki is not well established. Some trends in the data are suggestive but do not achieve statistical significance, and the ^{152}Eu analysis in particular is very strongly dependent on a few influential observations at the greatest distances. Showing the absence of an effect amounts to "proving a negative." The relative paucity of measurements in Nagasaki, particularly at greater distances, is problematic. To a great extent, this lack of longer-range measurements in Nagasaki is driven by the lower overall neutron fluences there. One cannot measure as far from the hypocenter in Nagasaki as in Hiroshima, for a given limit of detection, because the values of neutron activation overall are somewhat lower in Nagasaki.

The low-level measurement situation is different among the three main radionuclides that have been measured for thermal-neutron activation. For ^{36}Cl, the detection limit might be some 2 powers of 10 below the apparent natural background level in materials similar to the samples of interest, which presumably is due to cosmic-ray production over geological time. The natural background level has been measured in several types of relevant sample materials and appears to be reasonably consistent overall with the level that is approached in the deeper portions of large concrete cores. Nevertheless, the limit on detectability of ^{36}Cl attributable to the bomb fluence might prove to be determined by the uncertainty in the sample-specific level of ^{36}Cl due to cosmic-ray production, which has some uncertainty in addition to counting statistics. There is substantial potential variation among samples in the saturation level of the chlorine in the sample, which presumably is due to the geological and hydrological history of the chlorine involved. In concrete cores, although the material was presumably homogenized at the time of construction in very recent geological time, so that cosmic background should be homogeneous throughout a concrete pour, many cores are not deep enough to approach an unequivocal asymptote that clearly applies to the concrete in question. In fact, the situation is even more complicated, in that concrete can contain inclusions in the form of rocks, pebbles, and so on, that could dominate the material in a given slice of a core and have a different background level from the concrete itself.

Another source of measurement error in the case of ^{36}Cl is the possible dilution of the stable-chlorine pool by infiltration of rainwater into the sample matrix in situ, which needs further evaluation.

Several types of possible errors are related to the depth profile of ^{36}Cl activation in a concrete core: physical modifications of the exposed surfaces of concrete structures (adding or removing material), which might affect the depth profile near the surface or cause uncertainty in the effective depth of a slice from a core at the time of sampling versus ATB, and unexpected buildup or other unforeseen effects due to the interactions of incident high-energy neutrons near the exposed surface.

In addition, there is an issue of possible error due to production of ^{36}Cl from a competing neutron reaction on potassium. However, the classifications pertaining to depth profile and competing production by the ^{39}K$(n,\alpha)^{36}$Cl reaction are related to errors in calculated values rather than measured values.

For ^{60}Co, the background samples that have been measured have rather high detection limits, especially because of relatively small content of stable cobalt in the steel samples that were measured. The only available indications of likely natural background levels of this radionuclide in iron and steel come from calculation and from measurements in laboratory reagents that contain large masses of concentrated stable cobalt. These natural background indications are below the levels of interest in the more distant bomb samples by less than a power of 10, and the situation is degrading with each passing year because of the radioactive decay of the ^{60}Co from the bomb fluence.

Thus, natural background levels due to cosmic-ray neutrons appear to be a small but not negligible source of bias in the more distant ^{60}Co measurements. That effect has been evaluated here and an effort has been made to correct for it, but better in situ measurements of background in true environmental samples would be helpful.

All but perhaps one of the reported ^{60}Co measurements appear to lie above the nominal calculated MDCs reported here. (More detail on specific measurements is given in the body of Chapter 3.) If samples contained naturally occurring ^{60}Co at exactly 3.33×10^{-6} Bq/mg due to cosmic-ray production, this would increase the applicable background count rate by no more than 5%, except for the larger and more distant samples of Shizuma and others (1992), which would see increases up to about 14%. The actual increase in the MDC would depend on the assumed statistical distribution among samples of true values for ^{60}Co due to cosmic-ray production, but the effect would be minor for most cases of interest.

For ^{152}Eu, there are no relevant background samples at all. Distant samples have not been measured, and the deeper portions of granite cores are too small and too close to some portion of the surface exposed to the bomb fluence to provide an indication of natural background levels in the sample materials of interest. The sample from the basement of the A-bomb Dome building (Shizuma and others 1992) has a detection limit that is insufficient to measure the levels of interest. Again, the only available indications of likely natural background levels of this radionuclide in rocks, concrete, and ceramic tiles come from calculation and from measurements in laboratory reagents that contain large masses of concentrated stable europium. These natural background indications lie below the levels of interest in the more distant bomb samples by at least 2 powers of 10; this suggests that they should not be a significant source of bias in the measurements done to date. However, that has not been confirmed with true background samples of rocks, concrete, and ceramic tiles.

In the case of ^{152}Eu, some of the most distant measurements are close to or below the MDCs calculated here and should be interpreted with caution. That also implies that current methods might not have sufficient precision to provide useful estimates of natural background levels in the sample materials of interest; therefore, it might not be feasible to measure true background levels in sample-type materials. *However, it might still be of interest to extract and measure a few large background samples, in order to reduce the MDC for excluding natural background levels further below the range currently being reported in measurements.*

There is a possibility that measurements of ^{152}Eu at greater distances are affected by the counting interferences discussed above. It is not yet clear that all the measurements made by Nakanishi and others at 39–40 keV are free of potential bias from these sources. The method described by Shizuma and others (1993) of comparing the results of the 122 and 344-keV regions is intended to address this issue, but it might not have adequate sensitivity to identify all measurements that are affected by the counting interferences in the 122-keV region at these low levels. These issues deserve further attention and clarification.

Finally, substantial issues related to various aspects of quality assurance cannot be quantitatively evaluated here. Any future work should give serious consideration to maximizing the value of this important body of work by following a well-designed program of remeasurements and intercomparisons with stringent data-quality objectives.

SUMMARY AND CONCLUSION

Careful Analysis of the Measurement Data Has Results in Several Important Observations

The uncertainty used to characterize published measurements should be increased somewhat in most cases by calculating a total combined uncertainty for each measurement to account for all sources of random error that might have affected final reported values.

Statistical simulations indicate that the dispersion among the measurements, even after correcting some of the measurements for sample composition and local environment using the limited sample-specific calculated values that are currently available, is much too large to be consistent with the estimated uncertainties of the measurements, when those uncertainties are calculated based on propagation of error methods applied to the measurement process. Some of this apparent over-dispersion among the measurements could clearly be reduced by using detailed models of samples and their environs to create more accurate sample-specific calculated values for all of the measurements. However, some of this over-dispersion may also reflect sources of random error in the measurement process that are still unknown.

When the estimated uncertainties are increased to the extent that appears appropriate on the basis of the (admittedly sparse) information available for the present analysis, a discrepancy with DS86-calculated values clearly still remains in Hiroshima.

The data for Nagasaki are to some extent suggestive of a discrepancy, but more measurements are necessary to resolve this issue.

The discrepancy with DS86 in Hiroshima is statistically fairly robust and does not appear to be attributable solely to the less precise measurements made at the greatest distances.

When the measurements are fitted with a model that allows the relaxation length to vary, as should be allowed because of physical considerations, it appears that the initial relaxation length near the hypocenter is close to DS86 values, but the relaxation length increases more rapidly with distance than DS86. That might offer a clue to the nature of the discrepancy.

The ^{36}Cl measurements have several significant, recently discovered complications that must be resolved before they can be subjected to useful analysis.

The ^{60}Co and ^{152}Eu measurements are subject to several important concerns that could be addressed by a program of additional measurements or remeasurements and intercomparisons among laboratories.

Fitting a rapidly changing relaxation length involves an effect at the greater distances of interest in the Hiroshima neutron activation measurements (say, 1000–2000-m slant range) that is similar to fitting a finite asymptote, which might correspond to a "background effect" of some kind; and the two models might not be statistically distinguishable from each other in these data.

Appendix C

Cosmic-ray Neutron Contribution to Sample Activation

Cosmic rays that reach the Earth's atmosphere produce a variety of secondary particles from interactions with nitrogen, oxygen, and argon atoms. (NCRP 1987; UNSCEAR 1994; UNSCEAR 2000). The fluence of neutrons produced extends to very high energies (more than several gigaelectonvolts (Goldhagen and others 2000). The spectral distribution is relatively constant at atmospheric depths greater than a few hundred grams per square centimeter at geomagnetic latitude of 45 deg. N; the fluence rate is about 50% higher than previously reported measurements that probably did not account for the fluence of very high energy neutrons accurately (UNSCEAR 2000). The fraction of the total fluence that is below about 1 keV at ground level, and can thus slow down sufficiently in the sample to contribute to the thermal or resonance activation, is about 15–30%. The fraction can vary widely depending on local scattering and spallation effects.

The cosmic-ray neutron fluence is known to vary with geomagnetic latitude and with altitude because of the effect of the earth's magnetic field on the incident cosmic-ray particles. The total fluence is much lower (by about a factor of 2) near the equator than at the poles (UNSCEAR 1994). Estimates of the variations in total fluence with geomagnetic latitude are uncertain because of the scarcity of measurement data. UNSCEAR (1994) estimated that the total fluence in Tokyo was about half that at 45° N on the basis of a single measurement of dose-equivalent reported for Tokyo. However, Lal (1991) developed a polynomial fit to available low-energy neutron fluence measurements at various sites around the world that indicate that the fluence at sea level at 25° N (the geomagnetic latitude of Tokyo) should be about 75% of the value at 45° N.

The cosmic rays incident on the earth's atmosphere also vary with time because of variations in solar activity with an 11-year cycle. The time variation at sea level is fairly small—about 10% peak-to-peak (NCRP 1987)—and because the exposures of interest take place over an entire cycle or more, the temporal variations can generally be neglected.

Activations of environmental samples at any site will depend not only on the neutron fluence in air near the sample, but also on the amount of local scattering, buildup, and attenuation that will affect the fraction of the fluence that can cause activation in the sample (that is neutrons slowing down) and by any shielding of the sample by overlying material. Measurements in rock of ^{36}Cl activation (Linus and others 1999) indicate that the thermal activation first increases with depth to a depth of about 30–50 g/cm^2 and then decreases exponentially with a relaxation length of about 230 g/cm^2. The initial increase to about 30% over that at the surface is due to spallation reactions of the high-energy portion of the neutron fluence with atoms in the rock that produce showers of lower energy neutrons, i.e., an increase in the total low-energy neutron fluence. These high-energy nuclear reactions also occur in the atmosphere due to interactions of incident particles with atmospheric nuclei, producing a variety of cosmogenic radionuclides, including ^{10}Be, ^{26}Al, and even some ^{36}Cl (from reactions with argon (NCRP 1987)).

One can roughly estimate the amount of thermal activation that will occur in an unshielded sample from the estimated incident fluence, assuming that the sample has been exposed continuously to the same cosmic ray fluence for at least 3 or 4 half-lives of the activation product under consideration. If so, an equilibrium condition (saturation) will be achieved in which the rate of production of the decaying product will become equal to the production rate. Thus, the activity (A) at equilibrium will be given by

$$A = \varphi \sigma N$$

where φ is the effective thermal-neutron fluence, σ is the effective activation cross section, and N is the number of target atoms. $N = (N_{av}./W)f$, where f is the isotopic fraction of the target (natural abundance), W is the atomic weight of the target atom, and $N_{av}.$ is Avogadro's number. The appropriate cross section depends on the spectral distribution of thermal neutrons. Averaging over a 300°K Maxwellian distribution with a 1/E tail gives a weighted cross section about one-third lower than the published "thermal" cross section at 300°K (Kaul and others 1994) (see Table 3-1 of Chapter 3).

Several investigators attempted to calculate the cosmic-ray production of ^{60}Co and ^{152}Eu due to thermal neutron activation of ^{59}Co and ^{151}Eu. However, for the fluence term in the above equation, they used the older UNSCEAR (1994) estimate for high geomagnetic latitudes of 0.008 n/cm^2-s instead of the more recent estimate of 0.012 n/cm^2-s (Goldhagen and others 2000; UNSCEAR 2000). This value was an estimate of total fluence, however, and only incident thermal and epithermal neutrons (which slow to thermal in the sample) will activate ^{151}Eu and ^{59}Co (some ^{152}Eu and a substantial fraction of the ^{60}Co is formed by incident neutrons above thermal because of resonance in the region below 1 KeV (see Table 3-1 of Chapter 3). They also used the thermal-activation cross section at 300°K that might overestimate the activation (Kaul and others 1994). Because of the variations with geomagnetic latitude, the total neutron fluence at sea level in Japan is probably around 0.008 n/cm^2s^{-1} as was used, however, only a fraction of that will be converted to thermal energies in the sample.[1] Thus, one would expect that these calculated activation estimates could be too high by a factor of 2–3. If saturation had not been reached in the sample owing to its not being exposed to the estimated fluence for a long enough period, the true activity would be lower. Because of these local scattering and shielding effects, the uncertainty in any calculated value is very high; thus, a good estimate of cosmic-ray activation can be obtained only through measurement of environmental samples that have been exposed to cosmic radiation and are identical in almost all respects with the samples of interest (exposure time, shielding, materials, scattering, geomagnetic latitude, altitude, and so on) but have not been exposed to bomb neutrons. Because of its long half-life, the copper samples analyzed for ^{63}Ni would not probably have been exposed in situ for sufficient time for equilibrium to have been achieved.

An estimate of ^{35}Cl thermal activation by cosmic rays is even more dependent on background measurements because of the long half-life of ^{36}Cl (300,000 y). Because of that long half-life no natural chlorine samples have been exposed in situ for even a small fraction of their cosmic-ray exposure and most sources of chlorine present in concrete or granite originated from sources that have probably been heavily shielded from cosmic rays for all but the last 50,000–100,000 y. Thus, the fraction of saturation reached would generally be only a few percent of the equilibrium level and can be expected to vary widely, depending on the exposure history of the chlorine-containing materials that are present in the sample. The contributions to environmental ^{36}Cl from spallation reactions occurring continuously in the atmosphere and alternative production reactions (such as $^{39}K(n,\alpha)$), as well as the fact that the Earth's magnetic field has not remained constant over several hundred thousand years, make a true estimate of the possible level in any real sample suspect. Values of ^{36}Cl in sands and rocks in the Northern Hemisphere have been reported to range from less than 100×10^{-15} to 600×10^{-15} $^{36}Cl/Cl$ (Straume and others 1992). The estimated background levels of ^{36}Cl in the bomb samples analyzed by Straume and others (1992)—around $100–130 \times 10^{-15}$—are roughly the same as the reported values in rocks and sands in the Northern Hemisphere, accounting for variations with geomagnetic latitude.[2]

Except for ^{36}Cl, environmental background measurements of the cosmic-ray contribution to the reported activation measurements have not been performed.

---

[1] However, if the sample is surrounded by relatively high Z material, additional evaporation neutrons will be generated, as discussed by Linus and others (1999), and thus the thermal and epithermal fluence incident on the sample might be even higher.

[2] For example, the surface activity of ^{36}Cl in rock measured at an altitude of 2 km in the Sierra Nevada by Linus and others (1999) after using the model of Lal (1991) to correct to sea level and 25° geomagnetic latitude was about 300×10^{-15} $^{36}Cl/Cl$.

Shizuma attempted to measure ^{60}Co in a sample of steel obtained far from the epicenter (army storehouse), but the amount of [59]Co in the sample was too low to allow a reasonable measurement of cosmic-ray activity (see Appendix B-2). Measurements of the cosmic ray production of [63]Ni in copper at a church in Germany and in samples from other sites are in progress (Ruehm and others 2000a).

Shizuma (1999) and Komura (2000) have reported measurements of cosmic-ray activation in laboratory reagents. For ^{60}Co, Shizuma reported a value of $0.57 \pm 0.06 \times 10^{-6}$ Bq/mg of CoO, which corresponds to about 0.72×10^{-6} Bq/mg of cobalt; Komura reported a value of 3.5×10^{-6} Bq/mg of cobalt. For the europium reagent, Shizuma reported a value of $4.2 \pm 0.08 \times 10^{-6}$ Bq/mg Eu_2O_3, corresponding to about 5×10^{-6} Bq/mg of natural europium; Komura reported a value of 2.3×10^{-5}. In both cases, the values reported by Shizuma are about one-fifth of those reported by Komura. The large discrepancy might indicate that the Komura samples were exposed to a higher neutron fluence due to the amount and composition of shielding material surrounding the storage location. Shizuma suggested that the relatively low values he measured for [152]Eu might indicate that the reagent had not been exposed long enough to reach equilibrium. Using the above equation, and assuming that saturation had been reached, that the average thermal cross sections were 30 and 4400 barns for [59]Co and [151]Eu (see Table 3-1 of Chapter 3), and about 25% and 10% of the ^{60}Co and [152]Eu activation were from epithermal (resonance integral) neutrons, and that the natural [151]Eu isotopic fraction was 0.51, one can estimate the ^{60}Co and [151]Eu produced based on a cosmic-ray neutron fluence of 0.008 to be 3×10^{-6} Bq/mg for ^{60}Co and 7×10^{-5} Bq/mg for [152]Eu. The measurements by Shizuma thus correspond to an effective thermal plus epithermal fluence of about 2×10^{-3} from the ^{60}Co measurement and about 0.5×10^{-6} from the [152]Eu measurement, or about one-fourth and one-fifteenth of the total incident fluence. The estimated effective thermal fluence from the Komura [152]Eu data is about one-fourth of the expected total fluence. Thus, the Shizuma estimate of cosmic-ray ^{60}Co activation and the Komura estimate of [152]Eu activation are in reasonable agreement with what one might expect, whereas the Shizuma [152]Eu estimate appears to be too low and the Komura ^{60}Co estimate too high. However, because of the considerable uncertainty in the measurements and the expected variations in thermal fluence due to local scattering, spallation, absorbtion of thermal neutrons (such as boron in laboratory glassware), and attenuation—effects that are unknown for these reagent samples—we have adopted the Komura reagent values for [152]Eu (2.3×10^{-5}) and for [60]Co (3.5×10^{-6}), each with an estimated uncertainty of $\pm 25\%$ (1 SD). Note that the above equation suggests that the ratio of [152]Eu to ^{60}Co should be about 20–25 for cosmic rays, whereas both Shizuma's and Komura's measurements indicated a ratio of only about 8. A possible explanation is that the [152]Eu sample was not in equilibrium in either sample and the actual [152]Eu activation is therefore higher by a factor of 2–3 than the Komura measurement of about 6×10^{-5}.

It is interesting that Ichikawa (2000) reported results of measuring thermal neutrons using a [3]He detector at various locations in Hiroshima and Nagasaki. The values ranged from 123 to 200 counts/h in his detector, corresponding to an average thermal fluence near the ground of about 1.5×10^{-3} n/cm^2-s, about 20% of the estimated total fluence at Hiroshima given above. That is in reasonable agreement with the fraction of total fluence in the thermal range measured by Goldhagen and others (2000). However, the exact thermal fraction of the total is known to be highly sensitive to the local scattering medium and would be expected to be higher near soil, particularly wet soil, than far from the ground in higher Z material. It thus does not keep the total fluence at Hiroshima from being somewhat higher or lower than the 0.008 n/cm^{2}s^{-1} estimated, but it does indicate that the total cosmic-ray neutron fluence in Hiroshima is within the expected range.

Appendix D

Letter from Committee on Dosimetry to DOE

NATIONAL RESEARCH COUNCIL
COMMISSION ON LIFE SCIENCES
2101 Constitution Avenue Washington, D.C. 20418

BOARD ON RADIATION EFFECTS RESEARCH

NAS Room 342
TEL: (202) 334-2232
FAX: (202) 334-1639

August 26, 1996

Frank C. Hawkins, P.E.
Director, Office of International Health Programs (EH-63)
U S. Department of Energy
Washington, DC 20585

Dear Mr. Hawkins:

The National Research Council (NRC) Committee on Dosimetry for the Radiation Effects Research Foundation (RERF) is a small committee of approximately 6 members which was formed in 1988 to oversee the dosimetry activities associated with the RERF in Hiroshima and Nagasaki. Initially the committee was charged to oversee the ongoing uncertainty analysis and its documentation, to review the plans for the assessment of doses to factory workers and terrain-shielded survivors at Nagasaki, and to oversee the resolution of the difference between measured and calculated doses at Hiroshima. During the past 8 years, that committee has been studying the dosimetry research activities, dose classifications, and dose measurements relevant to the RERF sponsored by the U.S. Departments of Energy and

Defense. At the same time, Japan has had a dosimetry oversight committee which has been acting as an official U.S. counterpart. On occasion the U.S. and Japanese committees have met together to exchange information and assessments and to discuss future goals and experiments.

A joint meeting was held of the Committee on Dosimetry for the RERF and a Japanese Dosimetry Working Group on May 22–23, 1996 at the National Academy of Sciences' Beckman Center in Irvine, California. The NRC committee members included Rufard Alsmiller, Robert Christy, Alvin Weinberg, Wayne Lowder, Keran O'Brien, and me. Five representatives of the Japanese dosimetry working group (Soichiro Fujita, Masaharu Hoshi, Toshiso Kosako, Takashi Maruyama, and Kiyoshi Shizuma) made presentations and participated in the discussions. Also present were some members of the former U.S. working group on the DS86 dosimetry system who are still active in dosimetry work, including Dean Kaul, William Woolson, and Tore Straume. Additionally, representatives of U.S. DOE (Libby White), U.S. DOD-DNA (David Auton, John Bliss, and Robert Young), RERF (Dale Preston), and the Japanese Ministry of Health and Welfare (Hiroshi Maruyama) were present.

The meeting participants reviewed recent progress in A-bomb dosimetry work in the U.S. and in Japan and summarized the current status of the dosimetry. At the meeting, the joint working groups agreed upon a set of recommendations. The U.S. DOE representative present, Libby White, asked the NRC committee to write a brief letter report to the Office of International Health Programs summarizing the recommendations which were endorsed by the NRC committee.

As a preamble to the recommendations, I should emphasize that modern radiation protection is based on the risk coefficients for cancer derived from the A-bomb survivor study at Hiroshima and Nagasaki. The dosimetry of the survivors, which is used in the denominator in the risk coefficient, is as important as the assessment of radiation-induced cancers in the survivors. Consequently, the dosimetry must be studied until uncertainties in it can be reduced to a reasonable level. The uncertainty in the fast-neutron components at Hiroshima and Nagasaki, which are in doubt by perhaps a factor of 2 to 5 at Hiroshima, especially require urgent investigation. The urgency of the investigation is mandated by the fact that risk estimates are ongoing and epidemiology studies are constantly under revision. In addition, key scientists who have been studying RERF dosimetry are retiring, research teams are disbanding and facilities are losing their capability to conduct the needed studies due to a lack of funding, and copper wire from defined locations in the two cities needs to be located or it will be lost forever. To that end, the NRC committee makes the following recommendations:

1. That investigators vigorously pursue experiments that will lead to improved confidence in a revised DS86.
2. That investigations to resolve the neutron uncertainty be pursued, including
   • Evaluation (quality assurance) and intercomparison of U.S. and Japanese measurements of thermal neutrons in order to assess the handling of back-

ground problems (including the use of samples from long distances) and to assess total uncertainty in each measurement.

  • Application of the ^{63}Cu (np) ^{63}Ni reaction for fast-neutron measurements by both the U.S. and Japan (this requires an intensive search for copper samples, particularly up to 500m and beyond, if possible, in both Hiroshima and Nagasaki).

  • Calculations of weapon leakage and nitrogen cross-section experiments.

  3. That a revised DS86 include a re-evaluation of gamma rays at Hiroshima, yield, height of burst, the U.S. Army map (survivor locations), and shielding.

  4. That a strong effort be initiated to quantify uncertainties in all phases of DS86 and any later revision with a view to upgrading all estimates of uncertainty that are an integral part of the dosimetry system.

The specific recommendations include additional scientific work that needs to be done. While some of the recommendations are perhaps more appropriate to be made to DOD-DNA, two of the projects are specifically directed to DOE. The first of these, identified as recommendation 2, first bullet, involves the setting up of a small team composed of 2 knowledgeable investigators, 1 U.S. and 1 Japanese, to make a thorough examination of all measurements of neutron activation in the U.S. and Japan (cobalt, chlorine, and europium). Examination of these measurements by the U.S. and Japanese groups should pay special attention to the handling of background and the assessment of uncertainties (additional background measurements involving distant samples may be necessary). The aim is to put all of the measurements on a common basis, thereby permitting consistent appraisal and facilitating judgments about their relative significance (a quality assurance evaluation). The second recommendation (second bullet) concerns the development of the ^{63}Cu (np) ^{63}Ni reaction for fast-neutron measurement. Similar work will be done in Japan by Dr. Shibata but using a different assay system. The committee members believe that the conduct of these two projects is vital to our future knowledge of risk-estimation and the basis of radiation protection standards world-wide.

Yours sincerely,

Warren K. Sinclair, Ph.D.
Chairman
Committee on Dosimetry for RERF

cc: Paul Gilman
    John Zimbrick
    Paul Seligman
    Libby White
    Charles Arbanas
    Report Review Committee

Glossary

AA–Atomic Absorption—Spectroscopy utilizing the emission, absorption, or fluorescence of light at discrete wavelengths by atoms in a vaporized sample to determine the elemental composition of the sample.

absorbed dose—When ionizing radiation passes through matter, some of its energy is imparted to the matter. The amount of energy absorbed/per unit mass of irradiated material is called the absorbed dose. It is measured in gray or rad.

activation—The process of making a material radioactive by bombardment with neutrons, protons, or other nuclear particles.

AMS–Accelerator Mass Spectroscopy—A method that employs two sophisticated methodologies—a particle accelerator and a mass spectrometer—to provide in this application an estimate of the number of neutrons that a sample was exposed to at the time of the bombing.

APRF—Aberdeen Pulse Reactor Facility

ATB–At Time of Bomb—A designation to indicate that a particular event occurred at the time that the bombs were detonated in Hiroshima and Nagasaki, i.e., a reference point in time.

ATM–At Time of Measurement—A designation to indicate that a particular event occurred at the time that a specific measurement was made (usually later than ATB).

attenuation—The process by which a beam of radiation is reduced in intensity when passing through some material. It is the combination of absorption and scattering processes that leads to a decrease in flux density of the beam when projected through matter.

background radiation (measurement/level)—The level, often low, at which some substance, agent, or event is present or occurs at a particular location and time in the absence of the radiation source under study.

biological dosimetry—The use of measurements in biological samples to provide an estimate of radiation exposures; examples include measurement of chromosome aberrations in blood cells and measurement of thermoluminescence emitted from tooth enamel.

chromosome aberrations—Alteration from normal structure or number of chromosomes. i.e. dicentrics, translocations, etc.

Co–cobalt—Element number 27 of the periodic table. Isotopes such as ^{60}Co can be formed by neutrons irradiating ^{59}Co(n,γ) often found in steel.

cross-section—A measure of the probability that a nuclear reaction will occur. Usually measured in barns. It is the apparent or effective area presented by a target nucleus, or particle, to an oncoming particle or other nuclear radiation, such as a neutron.

CV–Coefficient of Variation—A measure of relative variation expressed by the standard deviation as a percentage of the mean.

DDREF—Dose and Dose-rate Effectiveness Factor—A factor used to adjust for the different biological effect with different doses and dose rates of low-LET radiation from those at which the original data was obtained. Typically, the factor refers to the possible reduction in carcinogenesis at low doses and/or at low dose rates.

delayed neutrons—During the fission process in the detonation of the atomic bomb, some neutrons are emitted immediately as "prompt" neutrons, while others are emitted after a very short time period and are referred to as "delayed neutrons."

dicentrics—A type of chromosome aberration visible through the light microscope in which two chromosomes with broken ends rejoin to form a chromosome with two centromeres.

DNA–Defense Nuclear Agency—A division of the U.S. Department of Defense that currently has been replaced by the Defense Threat Reduction Agency (DTRA).

dose—A term denoting the amount of energy absorbed from radiation.

down-scattering—The loss of energy by neutrons undergoing elastic collisions with the nuclei of the atmosphere and other intervening material.

DS86—The designation of the dosimetry system that was adopted in 1986 and is currently used to express doses to A-bomb survivors. It is also used by the Radiation Effects Research Foundation in the assessment of risk following exposures to the radiation from the atomic bombs in Hiroshima and Nagasaki.

EAR–Excess Absolute Risk—The excess risk attributed to irradiation and usually expressed as the numerical difference between irradiated and non-irradiated populations (e.g., 1 excess case of cancer/1 million people irradiated annually for each gray). Absolute risk may be given on an annual basis or lifetime (70-yr) basis.

electron spin resonance (ESR)—The measurement of magnetic resonance arising from the magnetic movement of unpaired electrons in a paramagnetic substance or in a paramagnetic center in a diamagnetic substance.

electron volt (eV)—a unit of particle energy.

EML–DOE Environmental Measurements Laboratory—A laboratory responsible for measurements of radiation that is located in New York City and operated by the U.S. Department of Energy.

energy bin—Radiation emitted from a source, such as an atomic bomb, can cover a broad energy spectrum. This spectrum can be divided into specific energy groupings called "energy bins."

epicenter—The point at which the detonation of the atomic bomb actually occurred.

epidemiology—The study of diseases as they affect populations, including the distribution of disease, or other health-related states and events in human populations; the factors (e.g., age, sex, occupation, economic status) that influence this distribution; and the application of this study to assess and control health risk.

epithermal neutrons—A neutron having an energy in the range immediately above the thermal range, roughly between 0.02 and 100,000 eV.

equivalent dose—A unit of biologically effective dose, defined by the ICRP in 1990 as the absorbed dose in gray multiplied by the radiation weighting factor (Wr). For all x-rays, gamma rays, beta particles, and positrons the radiation weighting factor is 1; for alpha particles it is 20; for neutrons it depends on energy (see ICRP 1991) and dose level.

ERR–Excess Relative Risk—A model that describes the risk imposed by exposures as a multiplicative increment of the excess disease risk above the background rate of disease.

Eu–europium—One member of the rare earth elements in the cerium subgroup, element number 63. Formed when certain atoms are bombarded with neutrons from atomic bombs

fast neutrons—Neutrons with energy greater than approximately 100,000 electron volts.

FISH–Fluorescence In Situ Hybridization—The use of DNA libraries derived specifically from particular chromosomes and conjugated with fluorescent molecules to generate reagents that cause distinctive fluorescence on individual chromosomes. Chromosomal aberrations involving the transfer of DNA from one chromosome to another (such as reciprocal translocations) can be detected using this "chromosome painting."

fluence—The number of radioactive particles, neutrons, or photons per unit cross-sectional area.

free-field value—Also sometimes called free-in-air value, is the fluence one would calculate at a given distance if there were no surrounding shielding material such as building walls, etc., to either attenuate or scatter radiation.

FSD–Fractional Standard Deviation—The fractional standard deviation of a random variable is equal to its standard deviation σ divided by its mean μ. This ratio has also been called the "coefficient of variation." In practice, σ and μ are replaced by estimators such as sample standard deviation and a sample mean, respectively. The FSD is often expressed as a percent. When a measurement is considered as a random variable with some associated probability distribution, the FSD quantifies the relative precision of the measurement.

gamma ray—The most penetrating radiation produced by radioactive decay, or in some other nuclear process. Gamma rays can be blocked only by dense material such as lead.

globe-data shielding—A shielding model in DS86 to calculate the radiation shielding for the approximately 4,500 survivors in Hiroshima and Nagasaki for whom nine parameter shielding was not appropriate because they were either heavily shielded by concrete buildings or in the open, but shielded by a house or terrain.

gray (Gy)—A unit used to describe the amount of energy that radiation deposits or is absorbed in tissue; 1 gray = 100 rad.

ground range—The distance from the point on the ground immediately below the point where the bomb was detonated (i.e., the hypocenter) to a specific location; as opposed to a "slant range," which is the direct distance from the actual point of detonation (i.e., the epicenter) above ground to the specific point of interest.

half-life—The time it takes for a radioactive quantity to decrease by half; used to describe how long radioactive isotopes take to decay to half their original activity.

height of burst (HOB)—The height above the Earth's surface at which a bomb is detonated in the air.

hypocenter (ground zero)—The point on the Earth's surface vertically below or above the center of a burst of a nuclear (or atomic) weapon; frequently abbreviated to GZ.

in situ—In the original location. Any test conducted in the field or in a material such as granite, tissue, or cells.

ionizing radiation—Radiation that has sufficient energy to be capable of ionizing atoms or molecules.

kerma—Kinetic energy released in material. A quantity that represents the kinetic energy transferred to charged particles by the uncharged particles per unit of mass of a material.

La-lanthanum—Element number 57 of the periodic chart.

LANL-Los Alamo National Laboratory—A national laboratory in Los Alamos, NM, that played an important role in the development and testing of the Hiroshima and Nagasaki bombs.

late gamma rays—Gamma rays emitted in a delayed manner from the detonation of the atomic bomb.

linear dose response—A description of the response of a particular effect that is linear with radiation dose.

LLNL-Lawrence Livermore National Laboratory—A national laboratory in Livermore, CA, that played an important role in the development and testing of the Hiroshima and Nagasaki bombs.

Lu-lutetium—Element number 71 of the periodic chart.

MDA–Minimum Detectable Activity—The lowest radioactivity at which detection is possible and measurements above background can be obtained.

MDC–Minimum Detectable Concentration—The lowest concentration at which detection is possible and measurements above background levels can be obtained.

MHW-Ministry of Health and Welfare—The agency in the Japanese government that has been responsible for sharing the funding of the Radiation Effects Research Foundation with U.S. DOE and some of the work of the dosimetry working groups; currently called the Ministry of Health, Labour, and Welfare (MHLW).

Monte Carlo analysis—The computation of a probability distribution of consequences by means of a random sampling method analogous to the game of roulette. Combinations of events and outcomes that yield possible consequences are randomly selected according to a specified probability distribution. The resulting consequences are counted and used to estimate other probability distributions.

NAA–Neutron Activation Analysis—Activation analysis in which the specimen is bombarded with neutrons; identification is made by measuring the resulting radioisotopes.

NCRP-National Council on Radiation Protection and Measurements—A body chartered by the U.S. government to provide advice and to help solve the nation's radiation problems. It produces reports on selected aspects of radiation protection.

nine-parameter shielding—A model that uses nine discrete physical parameters to describe the shielding from radiation provided to survivors by the houses in Hiroshima and Nagasaki

NIST-National Institute of Standards and Technology—A federal laboratory institute charged with maintaining expertise relative to measurements and standards.

ORELA-Oak Ridge Electron Linear Accelerator—An accelerator in Oak Ridge, TN, used to increase the energy of neutrons.

ORNL-Oak Ridge National Laboratory—A national laboratory in Oak Ridge, TN, that has played a major role in the production of nuclear weapons and studying the biological and environmental effects of radiation.

prompt gamma rays—Gamma rays emitted within a time too short for measurement.

prompt neutrons—A neutron released coincident with the fission process, as opposed to neutrons subsequently released.

radiation protection (shielding)—Reduction of radiation by interposing a shield of absorbing material between any radioactive source and a person, work area, or radiation-sensitive device.

RBE–Relative Biological Effectiveness—Ratio of the biological effectiveness of one radiation, e.g., neutrons, to another, e.g., gamma rays. Strictly, RBE is not the ratio of effects but rather the ratio of absorbed doses to produce the same level of effect (NCRP 1990).

RERF–Radiation Effects Research Foundation—A binational research foundation located in Hiroshima and Nagasaki, Japan, sponsored by Japan and the U.S., and studying the health effects of the atomic bombs on the survivors of the two bombs (formerly the Atomic Bomb Casualty Commission—ABCC).

RL–Relaxation Length—The distance an exponentially decreasing function is reduced to 1/e of its original value, i.e., if $A = A' \exp(-x/RL)$, when $x = RL$, $A = A'/e$.

risk—The probability that harm, such as a fatal cancer, will occur.

SAIC–Science Applications International Corp.—A company that has been involved in complex calculations related to atomic-bomb dosimetry and DS86.

sievert (Sv)—The SI unit of equivalent dose or effective dose equal to the dose in grays multiplied by the radiation weighting factor of the radiation.

slant range—The distance from a given location, usually on the Earth's surface, to the point at which the explosion occurred.

S_n–basic discrete ordinates—S_n is the angular segmentation method used for writing the transport difference equations in a form suitable for computer calculations in the discrete-ordinates method of calculating radiation transport. (The n in S_n is the number of solid angle segments representing the polar angles in a cylindrical geometry or the total number of angles with nonzero weights in a spherical one-dimensional geometry.) A discrete representation of the spatial and energy variables in the discrete ordinates transport equation is obtained by dividing the geometry systems into a fine space mesh and by using a multigroup set of cross sections (Roesch 1987)). The principal feature in discrete-ordinates methods is the discrete representation of angular, energy, and spatial variables in the Boltzmann transport equation.

source term—(a) The amount of radionuclides or chemicals released from a site to the environment over a specific period for use in dose reconstruction. (b) The nature, energies, and amounts of the radiation released from a nuclear weapon.

spallation—Splitting of the nucleus of an atom by high-energy bombardment.

thermal neutrons—Neutrons in thermal equilibrium with their surrounding medium. Thermal neutrons are those that have been slowed down by a moderator to an average speed of about 2200 m/sec at room temperature from the much higher initial speeds they had when expelled by fission. Their energies are less than 0.02 eV.

thermoluminescence—One of two principle methods of solid-state dosimetry for the measurement of integrated dose in certain crystalline natural materials such as quartz.

TLD–Thermoluminescent Dosimeter—Type of crystal used to monitor radiation exposure by emitting light, often used in a body, wrist, or ring badge. Must be processed in order to be "read."

translocation—A type of chromosome aberration involving the transfer of genetic material from one chromosome to another, nonhomologous chromosome. An exchange of genetic material between two chromosomes is referred to as reciprocal translocation.

tumorigenic—Any external influence capable of stimulating an increased growth or proliferation of abnormal cells, to form a tumor.

yield—The total effective energy released in a nuclear (or atomic) explosion. It is usually expressed in terms of the equivalent tonnage of TNT required to produce the same energy release in an explosion. The total energy yield is manifested as nuclear radiation, thermal radiation, and shock (blast) energy, the actual distribution being dependent upon the medium in which the explosion occurs.

References

Auxier. 1991. Presentation to the Committee on Dosimetry for the RERF. Irvine, CA.

Auxier. 1999. Dosimetry for the survivors of the nuclear bombings of Japan: The problem of neutrons in Hiroshima. Presentation to the Health Physics Society Meeting. June 27–July 1.

Boyer, K., W. Horowitz, and R. Albert. 1985. Interlaboratory variability in trace element analysis. Analytical Chemistry 57:454–459.

Carnes, B. A., D. Grahn, and J. F. Thomson. 1989. Dose-response modeling of life shortening in a retrospective analysis of the combined data from the JANUS program at Argonne National Laboratory. Radiat. Res. 119:39–56.

Chomentowski, M., A. M. Kellerer, and D. A. Pierce. 2000. Radiation dose dependences in the atomic bomb survivor cancer mortality data: a model-free visualization. Radiat. Res. 153:289–294.

Cierjacks, S., F. Hinterberger, D. Shmalz, D. Erbe, P. v. Rossen, and B. Leugers. 1980. High precision time-of-flight measurements of neutron resonance energies in carbon and oxygen between 3 and 30 MeV. Nucl. Inst. Meth. 169:185–198.

Clark, M. J., C. A. Laidlaw, S. C. Ryneveld, and M. I. Ward. 1996. Estimating sampling variance and local environmental heterogeneity for both known and estimated analytical variance. Chemosphere 32:1133–1151.

Covelli, V., V. Di Majo, M. Coppola, and S. Rebessi. 1989. The dose-response relationships for myeloid leukemia and malignant lymphoma in BC3F1 mice. Radiat. Res. 119:553–561.

Cullings, H. 2000. Summary of tabulated data from the original DS86 calculation files. Radiation Effects Research Foundation. Hiroshima, Japan.

Drigo, L., G. Tornielli, and G. Zannoni. 1976. Polarization in the $^{16}O(n,n)$ reaction. Nuov. Cim. A. 31:1–16.

Egbert, S. 1999. Personal communication.

Egbert, S. and D. Kaul. 2000. Local shielding effects. Presentation at US/Japan— joint dosimetry workshop. March 13–14. Hiroshima, Japan.

Fleming, S. J. and J. Thompson. 1970. Quartz as a heat-resistant dosimeter. Health Phys. 18:567–8.

Glasstone, S. and P. J. Dolan 1977. The Effects of Nuclear Weapons. 3rd ed. US Department of Defense and the US Department of Energy. Washington, DC.

Goldhagen, P., M. Reginatto, and F. Hajnal. 1996. Neutron Spectrum Measurements at Distances up to 2 Km from a Uranium Fission Source for Comparison with Transport Calculations. Proc. of the American Nuclear Society Topical Meeting on Advancements in Radiation Protection Measurements. April 21–25.

Goldhagen, P., M. Reginatto, T. Kniss, J. W. Wilson, R. C. Singleterry, I. W. Jones, and W. van Steveninck. 2000. Measurement of the energy spectrum of cosmic-ray induced neutrons aboard an ER-2 high-altitude airplane. Presented for publication in the Proceedings of the International Workshop on Neutron Spectroscopy in Science, Technology and Radiation Protection. June 4–8. Pisa, Italy.

Gritzner, M. L. and W. A. Woolson. 1987. Sulfur activation at Hiroshima. US-Japan joint reassessment of atomic bomb radiation dosimetry in Hiroshima and Nagasaki: Final report. Radiation Effects Research Foundation, Hiroshima, Japan. 283–292.

Grogler, N., F. G. Houtermans, and H. Stauffer. 1960. Uber die datierung von karamik und ziegel durch thermoluminescence. Helvetica Physica Acta 33:595–596.

Haberstock, G., J. Heingel, G. Korschinek, H. Morinaga, E. Nolte, U. Ratzinger, K. Kato, and M. Wolf. 1986. Accelerator mass spectrometry with fully stripped [36]Cl ions. Radiocarbon 28:204–210.

Hale, G. M., P. G. Young, M. B. Chadwick, and Z. P. Cohen. 1994. Analysis for nitrogen and oxygen in the Hiroshima dosimetry discrepancy study. Proceedings of the American Nuclear Society Shielding Conference. Houston, Texas.

Hamada, T. 1983a. Measurement of ^{32}P activity induced in sulfur in Hiroshima. US-Japan joint reassessment of atomic bomb radiation dosimetry in Hiroshima and Nagasaki: First report. Radiation Effects Research Foundation, Hiroshima, Japan: 45–56.

Hamada, T. 1983b. [32]P activity induced in sulfur in Hiroshima: Reevaluation of data by Yamasaki and Sugimoto. Second US-Japan joint reassessment of atomic bomb radiation dosimetry in Hiroshima and Nagasaki: Second report. Radiation Effects Research Foundation, Hiroshima, Japan: 52–55.

Hamada, T. 1987. Measurement of ^{32}P in sulfur. US-Japan joint reassessment of atomic bomb radiation dosimetry in Hiroshima and Nagasaki: Final report. Vol. 2. Radiation Effects Research Foundation, Hiroshima, Japan: 272–279.

Harvey, J. A., N. W. Larson, and D. C. Larson. 1992. Measurement of the nitrogen total cross section from 0.5 MeV to 50 MeV, and the analysis of the 433 keV resonance. Proceedings of a Conference on Nuclear Data for Science and Technology. May 13–17, 1991. Julich, Germany.

Hasai, H., K. Iwatani, K. Shizuma, M. Hoshi, K. Yokoro, S. Sawada, T. Kosako, Y. Morishima. 1987. ^{152}Eu depth profile of stone bridge pillar exposed to the Hiroshima atomic bomb; Data aquisition of ^{152}Eu activities for the analysis of fast neutrons. US-Japan joint reassessment of atomic bomb radiation dosimetry in Hiroshima and Nagasaki: vol. 2: 295–309. The Radiation Effects Research Foundation. Hiroshima, Japan.

Hashizume, T., T. Maruyama, A. Shiragai, E. Tanaka, M. Izawa, S. Kawamura, and S. Nagaoka. 1967. Estimation of the air dose from the atomic bombs in Hiroshima and Nagasaki. Health Phys. 13:149–161.

Haskell. 2000. Personal communication.

Higashimura, T., Y. Ichikawa, and T. Sidei. 1963. Dosimetry of atomic bomb radiation by thermoluminescence of roof tiles. Science 139:1284–1285.

Hoshi, M. and K. Kato. 1987. Data on neutrons in Hiroshima. US-Japan joint reassessment of atomic bomb radiation dosimetry in Hiroshima and Nagasaki: 252–255. The Radiation Effects Research Foundation. Hiroshima, Japan.

Hoshi, M., Y. Ichikawa, and T. Nagatomo. 1987. Thermoluminescence measurement of gamma rays at about 2000 m from the hypocenter. US-Japan joint reassessment of atomic bomb radiation dosimetry in Hiroshima and Nagasaki. The Radiation Effects Research Foundation. Hiroshima, Japan. Final Report vol. 2:149–152.

Hoshi, M., K. Yokoro, S. Sawada, K. Shizuma, K. Iwatani, H. Hasai, T. Oka, H. Morishima, and D. J. Brenner. 1989. ^{152}Eu activity induced by Hiroshima atomic bomb neutrons: comparison with the ^{32}P, ^{60}Co, and ^{152}Eu activities in cosimetry system 1986 (DS86). Health Phys. 57:831–837.

Hoshi, M., S. Endo, J. Takada, M. Ishikawa, Y. Nitta, K. Iwatani, T. Oka, S. Fujita, K. Shizuma, and H. Hasai. 1999. A crack model of the Hiroshima atomic bomb: explanation of the contradiction of "Dosimetry system 1986". Radiat. Res. (Tokyo):40 Suppl.:145–154.

Hubbell, J. H. 1982. Trends in radiation dosimetry. Int. J. Appl. Radiat. Isot. 33:1269–1290.

Ichikawa, Y. 1965. Dating of ancient ceramics by the thermoluminescent method. Bull. Inst. Chem. Res. 43:1–6.

Ichikawa. Y. 2000. Background measurement by counter: Presentation at US-Japan joint dosimetry workshop. March 13. Hiroshima, Japan.

Ichikawa, Y., T. Nagatomo, M. Hoshi, and S. Kondo. 1987. Thermoluminescence dosimetry of gamma rays from the Hiroshima atomic bomb at distances of 1.27 to 1.46 kilometers from the hypocenter. Health Phys. 52: 443–451.

ICRP. 1991. Recommendations of the International Commission on Radiological Protection adopted by the Commission in November 1990. Pub. #60. Annals of the ICRP. vol. 21, no.1–3.

Iimoto. 1999. Improved accuracy in the measurement of ^{152}Eu induced by atomic bomb neutrons in Nagasaki. Rad. Prot. Dos. 81:141–146.

Jablon, S. 1971. Atomic bomb radiation dose estimation at ABCC. Tech rpt. 23–71. ABCC, Hiroshima, Japan.

Kato, K., M. Habara, Y. Yoshizawa, U. Biebel, G. Haberstock, J. Heinzl, G. Korschinek, H. Morinaga, and E. Nolte. 1990. Accelerator mass spectrometry of ^{36}Cl produced by neutrons from the Hiroshima bomb. Int. J. Radiat. Biol. 58:661–672.

Kaul, D. 1999. Uncertainty Estimates for DS86 Dosimetry. Slide presentation to the Committee on Dosimetry for the RERF. Jan. 1999.

Kaul, D. 2000. Effect of technology changes at Hiroshima, presented at joint US-Japan dosimetry workshop. Slide presentation. March 13–14. Hiroshima, Japan.

Kaul, D. and S. Egbert. 1989. DS86 uncertainty and bias analysis. Draft report. Dec. 30 (rev. 1992).

Kaul, D. and S. Egbert. 1998. DS86, its roots, its present and its future. Slide presentation to the Committee on Dosimetry for the RERF. June. Washington, DC.

Kaul, D. and S. Egbert. 2000. Presentation to joint US/Japan workshop in Hiroshima. March 13–14. Hiroshima, Japan.

Kaul, D., W. A. Woolson, S. Egbert, and T. Straume. 1994. A brief summary of comparisons between the DS86 A-bomb survivor dosimetry system and in-situ measurements in light of new measurements, revised nuclear data and improved calculational methods. Proceedings of the 8th International Conference on Radiation Shielding.

Kellerer, A. M. and E. Nekolla. 1997. Neutron versus gamma-ray risk estimates. Inferences from the cancer incidence and mortality data in Hiroshima. Radiat. Environ. Biophys. 36:73–83.

Kellerer, A. M. and Walsh. 2001. Risk coefficient for fast neutrons with regard to solid cancer. Radiat. Res. (submitted).

Kerr, G., F. Dyer, J. Emery, J. Pace, R. Brodzinski, and J. Marcum. 1990. Activation of cobalt neutrons from the Hiroshima bomb. ORNL 6590. Oak Ridge National Laboratory. Oak Ridge, TN.

Kimura, T. 1993. Determination of specific activity of ^{60}Co- in steel samples exposed to the atomic bomb in Hiroshima. Radioisotopes 41:17–20.

Kimura, T., N. Takano, T. Iba, S. Fujita, T. Watanabe, T. Maruyama, and T. Hamada. 1990. Determination of specific activity of cobalt (^{60}Co/Co) in steel samples exposed to the atomic bomb in Hiroshima. J. Radiat. Res. (Tokyo) 31:207–213.

Kodama, Y., D. Dawel, N. Nakamura, D. Preston, T. Honda, M. Itoh, M. Nakano, K. Ohtaki, and A. A. Awa. 2001. Stable chromosome aberrations in atomic bomb survivors: Results from 25 years of investigation. Radiation Research Vol. 156, in press, with permission of authors.

Komura, K. 2000. Contribution on environmental neutrons for the production of ^{152}Eu, ^{154}Eu and ^{60}Co. Presentation at the US–Japan joint dosimetry workshop. March 13. Hiroshima, Japan.

Komura, K., and Yousef. 1998. Abstract from a presentation at the 41st meeting of the Japan Radiation Research Society. December 2–4. Nagasaki, Japan.

Kratochvil, B., D. Wallace, and J. Taylor. 1984. Sampling for chemical analysis. A review of fundamentals. Anal. Chem. 56:113r–129r.

Lafuma, J., D. Chomelevsky, J. Chameaud, M. Morin, R. Masse, and A. M. Kellerer. 1989. Lung carcinomas in Sprague-Dawley rats after exposure to low doses of radon daughters, fission neutrons, or gamma-rays. Radiat. Res. 118:230–245.

Lal, D 1991. Earth Planet. Sci. Lett. 104:424–439.

Lillie, R. A., L. Broadhead, J. V. Pace, and D. G. Cacusi. 1988. Sensitivity/uncertainty analysis for the Hiroshima and Nagasaki dosimetric reevaluation. Nucl. Sci. Eng. 100:105.

Linus, D., D. Elmore, S. Vogt, P. Sharma, M. Bourgeois, and A. Dunne. 1999. Erosin-corrected ages of quaternary geomorphic events using cosmogenic ^{36}Cl in rocks. http//:primelab.physics.purdue.edu/.

Loewe, W. E. 1984. Hiroshima and Nagasaki initial radiations: delayed neutron contributions and comparison of calculated and measured cobalt activation. Nucl. Tech. 68:311–318.

Maruyama, T. 1983. Reassessment of gamma-ray dose estimates from thermoluminescent yields in Hiroshima and Nagasaki. US-Japan joint reassessment of atomic bomb radiation dosimetry in Hiroshima and Nagasaki. First Report pp.122–137. Radiation Effects Research Foundation. Hiroshima, Japan.

Maruyama, T. 2000. Update on TLD measurements in Hiroshima. Presentation at a meeting of the Commitee on Dosimetry for the RERF. January 10–11. Irvine, CA.

Maruyama, T., and Y. Kuramoto. 1987. Reassessment of gamma ray doses using thermoluminescence measurements. US-Japan Joint Reassessment of Atomic Bomb Radiation Dosimetry in Hiroshima and Nagasaki: Final Report vol. 2. 113–124. The Radiation Research Effects Foundation. Hiroshima, Japan.

Miller, M. and J. Fitzgerald. 1991. How to assess overall data quality. Poll. Eng. 23:74–76.

Nagatomo, T., M. Hoshi, and Y. Ichikawa. 1995. Thermoluminescence dosimetry of the Hiroshima atomic-bomb gamma rays between 1.59 km and 1.63 km from the hypocenter. Health Phys. 69:556–559.

Nakamura, N. 1999. Report on biological dosimetry: delivered to the Committee on Dosimetry for the RERF. January. Irvine, CA.

Nakamura, N. 2000. ESR vs. chromosome aberrations. Presented at the US/Japan joint workshop on RERF dosimetry. March 13–14. Hiroshma, Japan.

Nakamura, N., C. Miyazawa, S. Sawada, M. Akiyama, and A. A. Awa. 1998. A close correlation between electron spin resonance (ESR) dosimetry from tooth enamel and cytogenetic dosimetry from lymphocytes of Hiroshima atomic-bomb survivors. Int. J. Radiat. Biol. 73:619–627.

Nakanishi, T., T. Imura, K. Komura, and M. Sakanoue. 1983. [152]Eu in Samples exposed to nuclear explosions at Hiroshima and Nagasaki. Nature 302: 132–134.

Nakanishi, T., H. Ohtani, R. Mizuochi, K. Miyaji, T. Yamamoto, K. Kobayashi, and T. Imanaka. 1991. Residual neutron-induced radionuclides in samples exposed to the nuclear explosion over Hiroshima: comparison of the measured values with calculated values. J. Radiat. Res. (Tokyo) 32 Suppl:69–82.

Nakanishi, T., K. Miwa, and R. Ohki. 1998. Specific radioactivity of [152]Eu in roof tiles exposed to atomic bomb radiation in Nagasaki. J. Radiat. Res. (Tokyo) 39:243–250.

NCRP (National Council on Radiation Protection). 1987. Exposure of the population in the United States and Canada from natural background radiation. National Council on Radiation Protection and Measurements. Report #94. Bethesda, MD.

NCRP (National Council on Radiation Protection). 1993. Risk estimates for radiation protection. National Council on Radiation Protection and Measurements; Report #115. Bethesda, MD.

NCRP (National Council on Radiation Protection). 1997. Uncertainties in fatal cancer risk estimates used in radiation protection. National Council on Radiation Protection and Measurements: Report #126. Bethesda, MD.

NRC (National Research Council). 1987. An assessment of the new dosimetry for A-bomb survivors. Panel on reassessment of A-bomb dosimetry. National Academy Press. Washington, DC.

NRC (National Research Council). 1990. Health effects of exposure to low levels of ionizing radiation—BEIR V. National Academy Press. Washington, DC.

NRC (National Research Council). 1996. Letter from the Committee on Dosimetry to Frank Hawkins of DOE. Committee on Dosimetry for the Radiation Effects Research Foundation (RERF). Board on Radiation Effects Research. National Research Council (see Appendix D of this report)

NRPB (National Radiological Protection Board). 1993. Estimates of late radiation risks to the UK Population. National Radiological Protection Board. Chilton, U.K.

NUREG 1507. 1995. Minimum detectable concentrations with typical radiation survey instruments for various contaminants and field conditions. http://techconf.llnl.gov/radcri/1507.html.

Okajima, S. and J. Miyajima. 1983. Measurement of neutron-induced [152]Eu radioactivity in Nagasaki. US/Japan joint workshop for reassessment of atomic bomb radiation dosimetry in Hiroshima and Nagasaki. First report. February. Hiroshima, Japan.

Okumura, H. and T. Shimasaki. 1997. Reassessment of atomic bomb neutron doses (Japanese). Fiscal year 1996 Report of research group on atomic bomb related symptoms. Hiroshima, Japan.

Pace, J. 1993. Personal communication.

Pass, B., A. E. Baranov, E. D. Kleshchenko, J. E. Aldrich, P. L. Scallion, and R. P. Gale. 1997. Collective biodosimetry as a dosimetric "gold standard": a study of three radiation accidents. Health Phys. 72:390–396.

Pierce, D. A. and D. L. Preston. 2000. Radiation-related cancer risks at low doses among atomic bomb survivors. Radiat. Res. 154:178–186.

Pierce. D. A., Y. Shimizu, D. L. Preston, M. Vaeth, and K. Mabuchi. 1996. Studies of the mortality of atomic bomb survivors. Report 12, Part I. Cancer: 1950–1990. Radiat. Res. 146:1–27.

Preston, D. L. 1999. Impact of RERF dosimetry on data analysis. Presentation to the Committee on Dosimetry for the RERF. January 1999. Irvine, CA.

Preston, D. L., D. A. Pierce, and M. Vaeth. 1993. Neutrons and radiation risk. A commentary. RERF Update no. 4.

Rhodes, R. 1995. The making of the atomic bomb. Simon and Shuster, New York, NY.

Rhodes, W. A., J. N. Barnes, and R. T. Santoro. 1994. An explanation of the Hiroshima activation dilemma in the Hiroshima dosimetry discrepancy study. Proceedings of the American Nuclear Society Shielding Conference. Houston, TX.

Roesch, W. 1987. US-Japan joint reassessment of atomic bomb radiation dosimetry in Hiroshima and Nagasaki—Final Report. Vol. 1&2. Radiation Effects Research Foundation, Hiroshima, Japan.

Rossi, H. H. and A. M. Kellerer. 1974. The validity of risk estimates of leukemia incidence based on Japanese data. Radiat. Res. 58:131–140.

Rossi, H. H., and C. W. Mays. 1978. Leukemia risk from neutrons. Health Phys. 34 353–360.

Rossi, H. H. and M. Zaider. 1990. Contribution of neutrons to the biological effects in Hiroshima. Health Phys. 58:645–647.

Rossi, H. H. and M. Zaider. 1996. Comment on the contribution of neutrons to the biological effect at Hiroshima. Radiat. Res. 146:590–593.

Rossi, H. H. and M. Zaider. 1997. Continuation of the discussion concerning neutron effects at Hiroshima. Radiat. Res. 147:269–270.

Ruehm. W. 2000. Personal communication.

Ruehm. W., T. Huber, K. Kato, and E. Nolte. 2000a. Measurement of ^{36}Cl at Munich—a status report. Ludwig-Maximilians University, Munich. Technical Report. November. Munich, Germany.

Ruehm, W., K. Knie, G. Rugel, A. A. Marchetti, T. Faestermann, C. Wallner, J. E. McAninch, T. Straume, and G. Korschinek. 2000b. Accelerator mass spectrometry of ^{63}Ni at the Munich Tandem Laboratory for estimating fast neutron fluences from the Hiroshima atomic bomb. Health Phys. 79:358–364.

Shibata, T. 2000. Presentation at US/Japan—joint dosimetry workshop. March 13. Hiroshima, Japan.

Shibata, T., M. Imamura, S. Shibata, Y. Uwamino, T. Ohkubo, S. Satoh, N. Nogawa, H. Hasai, K. Shizuma, K. Iwatani, M. Hoshi, and T. Oka. 1994. A method to estimate the fast neutron fluence for the Hiroshima atomic bomb. J. Phys. Soc. (Japan) 63: 3546–3547.

Shizuma, K. 1997. Identification of ^{63}Ni and ^{60}Co produced in a steel sample by thermal neutrons from the Hiroshima atomic bomb. Nuc. Inst. and Meth. A 384:375–379.

Shizuma, K., 1998. Residual ^{152}Eu and ^{60}Co activity induced by atomic bomb neutrons in Nagasaki. Manuscript submitted for publication.

Shizuma, K. 1999. Contribution of background neutron activation in the residual activity measurement and present status of ^{152}Eu measurement for Nagasaki samples. Presentation at a joint meeting of the US and Japan working groups. Irvine, CA.

Shizuma, K. 2000a. ^{152}Eu Measurements (review and Nagasaki Data); presentation at US/Japan-joint dosimetry workshop. March 13. Hiroshima, Japan.

Shizuma, K. 2000b. Personal communication.

Shizuma, K., K. Iwatani, H. Hashi, T. Oka, H. Morishima, and M. Hoshi. 1992. Specific activities of ^{60}Co and ^{152}Eu in samples collected from the Atomic-Bomb Dome in Hiroshima. J. Radiat. Res. (Tokyo) 33:151–162.

Shizuma, K., K. Iwatani, H. Hasai, M. Hoshi, T. Oka, and H. Morishima. 1993. Residual ^{152}Eu and ^{60}Co activities induced by neutrons from the Hiroshima atomic bomb. Health Phys. 65:272–282.

Shizuma, K., K. Iwatani, H. Hasai, M. Hoshi, and T. Oka. 1997. ^{152}Eu depth profiles in granite and concrete cores exposed to the Hiroshima atomic bomb. Health Phys. 72:848–855.

Shizuma, K., K. Iwatani, H. Hasai, T. Oka, S. Endo, J. Takada, M. Hoshi, S. Fujita, T. Watanabe, and T. Imanaka. 1998. Residual ^{60}Co activity in steel samples exposed to the Hiroshima atomic-bomb neutrons. Health Phys. 75:278–284.

Shimizu, S. and T. Saigusa. 1987. Estimation of ^{32}P induced in sulfur in utility pole insulators at the time of the Hiroshima atomic bomb. US-Japan joint reassessment of atomic bomb radiation dosimetry in Hiroshima and Nagasaki: Final report. vol. 2. Radiation Effects Research Foundation, Hiroshima, Japan: 266–268.

Shimizu, Y., D. A. Pierce, D. L. Preston, and K. Mabuchi. 1999. Studies of the mortality of atomic bomb survivors. Report 12, part II. Noncancer mortality: 1950–1990. Radiat. Res. 152:374–389.

Sposto, R., D. O. Stram, and A. A. Awa. 1991. An estimate of the magnitude of random errors in the DS86 dosimetry from data on chromosome aberrations and severe epilation. Radiat. Res. 128:157–169.

Stram, D. O., R. Sposto, D. Preston, S. Abrahamson, T. Honda, and A. A. Awa. 1993. Stable chromosome aberrations among A-bomb survivors: an update. Radiat. Res. 136:29–36.

Straume, T. 1988. Research proposal to the Defense Nuclear Agency.

Straume, T. 2000a. Personal communication.

Straume, T. 2000b. Presentation at US/Japan joint dosimetry workshop. March 13–14. Hiroshima, Japan.

Straume, T., and J. N. Lucas. 1995. Validation studies for monitoring of workers using molecular cytogenesis. Biomarkers in Occupational Health. Joseph Henry Press.

Straume, T., S. D. Egbert, W. A. Woolson, R. C. Finkel, P. W. Kubik, H. E. Gove, P. Sharma, and M. Hoshi. 1992. Neutron discrepancies in the DS86 Hiroshima dosimetry system. Health Phys. 63:421–426.

Straume, T., L. J. Harris, A. A. Marchetti, and S. D. Egbert. 1994. Neutrons confirmed in Nagasaki and at the Army Pulsed Radiation Facility: implications for Hiroshima. Radiat. Res. 138:193–200.

Straume, T., A. A. Marchetti, and J. E. McAninch. 1996. New analytical capability may provide solution to the neutron dosimetry problem in Hiroshima. Radiat. Prot. Dosim. 67:5–8.

Straume, T, A. Marchetti., J. McAninch, W. Ruehm, and G. Korschinek. 2000. Progress report to the Committee on Dosimetry for the RERF.

Tatsumi-Miyajima, J. 1991. Physical dosimetry at Nagasaki—^{152}Eu of stone embankment and electron spin resonance of teeth from atomic bomb survivors. J. Radiat. Res. Suppl.: 83–98.

Thompson, D. E., K. Mabuchi, E. Ron, M. Soda, M. Tokunaga, S. Ochikubo, S. Sugimoto, T. Ikeda, M. Terasaki, S. Izumi, and D. L. Preston. 1994. Cancer incidence in atomic bomb survivors. Part II: Solid tumors, 1958–1987. Radiat. Res. 137:S17–67.

Tindel, S. 2000. Data Quality Objectives (DQO). http://www.hanford.gov/dqo/index.html.

UNSCEAR (United Nations Scientific Committee on the Effects of Atomic Radiation). 1988. Sources, Risks, and Effects of Ionizing Radiation. Report to the General Assembly, with annexes. United Nations Press. New York, NY.

UNSCEAR (United Nations Scientific Committee on the Effects of Atomic Radiation). 1994. Sources and Effects of Ionizing Radiation. United Nations Press. New York, NY.

UNSCEAR (United Nations Scientific Committee on the Effects of Atomic Radiation). 2000. Sources and Effects of Ionizing Radiation. United Nations Press. New York, NY.

US EPA. 1994a. Guidance for Data Quality Assessment. US Environmental Protection Agency. Washington, DC.

US EPA. 1994b. Guidance for the Data Quality Objectives Process. US Environmental Protection Agency. Washington, DC.

Whalen, P. P. 1994. Source replica calculations in the Hiroshima discrepancy study. Proceeding of the American Nuclear Society Shielding Conference. Houston, TX.

Wolf, C., J. Lafuma, R. Mass, M. Morin, and A. M. Kellerer. 2000. Neutron RBE for tumors with high lethality in Sprague-Dawley rats. Radiat. Res. 154: 412–420.

Woolson. 1993. DNA/RARP A-bomb Dosimetry Working Group: The neutron discrepancy findings and recommendations. Briefing presented to the Committee on Dosimetry for the RERF. October 14.

Yamasaki, F. and A. Sugimoto. 1987. Radioactive ^{32}P produced in sulfur in Hiroshima. US-Japan joint reassessment of atomic bomb radiation dosimetry in Hiroshima and Nagasaki: Final report vol. 2. Radiation Effects Research Foundation, Hiroshima, Japan: 246–247.

Biosketches

WARREN K. SINCLAIR, PhD, is president emeritus of the National Council on Radiation Protection and Measurements. Dr. Sinclair received his BSc and MSc from the University of New Zealand and his PhD from the University of London. His scientific interests include radiation protection, radiation physics, and radiation-risk estimation. Dr. Sinclair has served as professor of physics and of zoology at the University of Texas and as senior biophysicist and associate laboratory director at Argonne National Laboratory; he is currently professor emeritus at the University of Chicago. He has served on the International Commission on Radiation Units and on the International Commission on Radiological Protection (ICRP), and is an Emeritus Member of ICRP. He is the alternate delegate to the UN Scientific Committee on the Effects of Atomic Radiation and serves on committees of the IAEA and the CEC and on the board of the RERF. He has been a member of many national committees for the Department of Energy, National Aeronautic and Space Administration, National Institutes of Health, Veterans Administraton and the National Research Council, where he also served as chair of the Board on Radiation Effects Research. He is a member of numerous professional associations, including the Radiation Research Society (former president), the Society of Nuclear Medicine, the Society for Risk Analysis, the American Association of Physicists in Medicine (former president), the Health Physics Society, the Society of Nuclear Medicine, and the Radiological Society of North America.

HAROLD M. AGNEW, PhD, is the former director of Los Alamos Scientific Laboratory (now Los Alamos National Laboratory) and the retired president of General Atomics in San Diego. He received his AB from the University of Denver and his MS and PhD from the University of Chicago. His research interests include nuclear physics and its application to defense, energy, and the biological sciences. Dr. Agnew served as the scientific adviser to the Supreme Allied Commander Europe (NATO) and as chairman of the General Advisory Committee to the Arms Control and Disarmament Agency. He has served on the White House Science Council and is a member of the Council on Foreign Relations. He is a member of Sigma Xi and a fellow of the American Association for the Advancement of Science and a fellow of the American Physical Society. Dr. Agnew was elected to the National Academy of Engineering in 1976 and the National Academy of Sciences in 1979. He received the E. O. Lawrence Award from the Atomic Energy Commission and the Enrico Fermi Award from the Department of Energy. Dr. Agnew served two terms in the New Mexico State Senate as the first senator from Los Alamos County.

HAROLD L. BECK, BS, retired in 1999 as the director of the Environmental Sciences Division of the Department of Energy (DOE) Environmental Measurements Laboratory (EML) in New York City. Mr. Beck previously served as the director of the EML Instrumentation Division and as acting deputy director of the laboratory. Mr. Beck received a BS degree from the University of Miami summa cum laude and did graduate work in physics and mathematics at Cornell University from 1960 to 1962. He is the author or coauthor of over 100 publications in radiation physics, radiation protection, environmental radiation, dosimetry, and instrumentation. His development of the scientific approach to reconstructing fallout doses to the US population from above-ground nuclear-weapons tests in Nevada earned him the DOE Meritorious Service Award in 1988, the second highest award in the department. Mr. Beck is the scientific vice-president of the National Council on Radiation Protection and Measurements (NCRP) for radiation measurement and the chair of NCRP Scientific Committee 93, on radiation measurement. He also chaired Scientific Committee 64-20, on surface soil contamination. Mr. Beck was a US delegate to the International Electrotechnical Commission's Scientific Committee 45B, on radiation protection instrumentation. He is a member of the American Association for the Advancement of Science, the American Nuclear Society, and the Institute of Electronic and Electrical Engineers, and he is a fellow of the Health Physics Society.

ROBERT F. CHRISTY, PhD, is a professor emeritus at the California Institute of Technology. Dr. Christy received his BA and his MA from the University of British Columbia and his PhD from the University of California. His research interests include the effects of cosmic rays, astrophysics, and nuclear physics. Dr. Christy has held several academic positions throughout his distinguished career and, most notably, served the California Institute of Technology in several capacities over the last fifty years. Dr. Christy is a member of the National Academy of Sciences, the International Astronomers Union, the American Physical Society, and the American Astronomers Society.

SUE B. CLARK, PhD is a Meyer Distinguished Professor in the College of Sciences and an Associate Professor of Chemistry at Washington State University in Pullman, Washington. Her current research areas include the environmental chemistry of plutonium and other actinides, chemistry of high level radioactive waste systems, and chemistry of actinide bearing solid phases in natural environments. Her research efforts are supported by grants from the U.S. Department of Energy's Environmental Management Sciences Program, Natural and Accelerated Biological Remediation Program, Nuclear Education and Energy Research Program, Basic Energy Sciences Program, and contracts from British Nuclear Fuels Inc., as well as organizations at the Hanford and Idaho National Engineering and Environmental Laboratory sites. She holds a BS degree from Lander College (Greenwood, SC) and MS and PhD degrees in chemistry from Florida State University (Tallahassee, FL). Prior to joining Washington State University in 1996, she was an assistant research ecologist at the University of Georgia's Savannah River Ecology Laboratory (1992–1996), and senior scientist at Westinghouse Savannah River Company's Savannah River Technology Center (1989–1992). She is involved in numerous service activities; she serves on various committees for the National Research Council's Board on Radioactive Waste Management, as well as the Committee on Dosimetry for the Radiation Effects Research Foundation. She has received several awards, including the Young Faculty Achievement Award in the College of Sciences at Washington State University (1998–1999), a Young Investigator Award, National Academy of Sciences Program on Nuclear Accidents and Radioactive Contamination (1993–1994), and the George Westinghouse Signature Award of Excellence, Westinghouse Corporation (1991). She is a member of the American Chemical Society and Sigma Xi, the Scientific Research Society.

NAOMI HARLEY, PhD, is a research professor of environmental medicine at New York University (NYU) School of Medicine. Dr. Harley received her PhD in radiological physics from NYU, her ME in nuclear engineering from New York University, her BE in electrical engineering from The Cooper Union, and an APC in management from the NYU Graduate Business School. Dr. Harley serves on the Medical Isotopes Committee at NYU Medical School, she is a Council Member in the National Council on Radiation Protection and Measurements (NCRP), and is an adviser to the US Delegation of the UN Committee on the Effects of Atomic Radiation. She is a member of the Editorial Board of the journal *Environment International.* She was elected a fellow in the Health Physics Society. She has published over 100 journal articles and chapters in six books, and she holds three patents at NYU for radiation detection devices. She chairs an NCRP committee on the health effects of radon and is the president of the Radon Section in the Health Physics Society. Her current research involves the dosimetry of internally deposited radionuclides, the measurement of radiation and radioactivity, and risk modeling of radiation carcinogenesis.

ALBRECHT KELLERER, PhD, is the director of the Radiobiological Institute of the University of Munich and of the Institute of Radiation Biology of the Forschungszentrum für Umwelt und Gesundhelt (GSF), National Research Center for Environment and Health. He was formerly professor of radiation biophysics at Columbia University, and later professor and chief of the institute for medical radiation research at the University of Wurzburg. Dr. Kellerer's research specialties include microdosimetry, radiation-risk assessment, and radiobiology. He is a member of the German National Commission for Radiation Protection, chairman of its committee for risk assessment, and a member of committees of International Commission of Radiation Units and measurements and International Council for Radiation Protection. Dr. Kellerer is the managing editor of the *Journal of Radiation and Environmental Physics*.

KENNETH J. KOPECKY, PhD, is a member of the Division of Public Health Sciences of the Fred Hutchinson Cancer Research Center and an affiliate professor in the Department of Biostatistics at the University of Washington in Seattle, Washington. He earned his PhD in statistics from Oregon State University. He was formerly a research associate at the Radiation Effects Research Foundation in Hiroshima, Japan, and a member of the Technical Steering Panel for the Hanford Environmental Dose Reconstruction Project. He is a coinvestigator in the Hanford Thyroid Disease Study and in the Fred Hutchinson Cancer Research Center—Russian project of the International Consortium for Research on the Health Effects of Radiation.

WAYNE M. LOWDER, AB, received his degree in physics from Harvard University and did graduate work in physics at Columbia University and the International School of Nuclear Science and Engineering, Argonne National Laboratory. He was physicist and later director of the Radiation Physics Division, Environmental Measurements Laboratory, Department of Energy (DOE), before his retirement in 1994. His primary fields of research were the physical properties, measurement, and dosimetry of natural and human-made radiation and radionuclides in the environment. He co-organized the five international symposiums on the natural radiation environment (1963–1995) and served on committees of the National Council on Radiation Protection and Measurements and the International Commission on Radiological Protection and in the secretariat of the UN Scientific Committee on the Effects of Atomic Radiation. From 1978–1986, he assisted the DOE Office of Health and Environmental Research in developing research programs in radiation measurement and dosimetry and on environmental radon. Prior to his service on the Committee on Dosimetry for RERF, he was the DOE representative to the National Research Council committee chaired by Frederick Seitz that oversaw the development of DS86.

ALVIN M. WEINBERG, PhD, is a distinguished fellow with the Oak Ridge Associated Universities. Dr. Weinberg is a member of the National Academy of Sciences and the National Academy of Engineering. He has performed extensive research on the design, development, and safety of nuclear reactors and is knowledgeable in risk assessment and analysis.

ROBERT W. YOUNG, PhD, is a radiation health-effects consultant with special interests in biological effects of neutrons, dose-determination methods for the definition of human health effects, and modeling of the physiological consequences of exposure to ionizing radiation. During the last thirty years, he served as director of the Biomedical Research Program at the Defense Nuclear Agency and as a division head at the Armed Forces Radiobiology Research Institute. During that time, he developed and directed programs in biological dosimetry, modeling of physiological effects of ionizing radiation, and effects of ionizing radiation, especially neutrons, on the nervous system. Dr. Young has chaired the NATO Project Group on Radiation Anti-emetic Drugs, served on committees for the International Atomic Energy Agency, provided expert advice to the US Nuclear Regulatory Commission, and served on the White House committee on the Chernobyl nuclear accident. He earned his MA and PhD at the Catholic University of America. Dr. Young has a long-standing interest in dose-determination methods for Hiroshima and Nagasaki and has served the National Research Council Committee on Dosimetry for the RERF as technical adviser and as ad hoc head of the Technical Working Group on Hiroshima neutron calculation. His current interest is in dose reconstruction for epidemiologic studies of radiation risk.

MARCO ZAIDER, PhD, is attending physician and head of brachytherapy physics at Memorial Sloan-Kettering Cancer Center and professor of physics (in radiology) at Cornell Medical School. He also served as a professor of clinical radiation oncology, public health, and applied physics in the department of applied physics at Columbia University. Dr. Zaider is also the director of the graduate program in medical physics at Columbia University. He received his MSc from the University of Bucharest and his PhD from the University of Tel Aviv. Dr. Zaider's scientific interests include radiation biophysics, microdosimetry, and medical physics. Dr. Zaider is the coauthor of *Microdosimetry and Its Applications*.